Science

Gray Wolf

Harcourt
SCHOOL PUBLISHERS

Orlando   Austin   New York   San Diego   Toronto   London

Visit *The Learning Site!*
www.harcourtschool.com

Copyright © 2006 by Harcourt, Inc.

All rights reserved. No part of this publication may be reproduced or transmitted in any form or by any means, electronic or mechanical, including photocopy, recording, or any information storage and retrieval system, without permission in writing from the publisher.

Requests for permission to make copies of any part of the work should be addressed to School Permissions and Copyrights, Harcourt, Inc., 6277 Sea Harbor Drive, Orlando, Florida 32887-6777. Fax: 407-345-2418.

HARCOURT and the Harcourt Logo are trademarks of Harcourt, Inc., registered in the United States of America and/or other jurisdictions.

Printed in the United States of America

ISBN 0-15-343586-0

1 2 3 4 5 6 7 8 9 10 032 14 13 12 11 10 09 08 07 06 05

Gray Wolf

Consulting Authors

Michael J. Bell
*Assistant Professor of Early
   Childhood Education*
College of Education
West Chester University of
   Pennsylvania

Michael A. DiSpezio
Curriculum Architect
JASON Academy
Cape Cod, Massachusetts

Marjorie Frank
Former Adjunct, Science Education
Hunter College
New York, New York

Gerald H. Krockover
*Professor of Earth and Atmospheric
   Science Education*
Purdue University
West Lafayette, Indiana

Joyce C. McLeod
Adjunct Professor
Rollins College
Winter Park, Florida

Barbara ten Brink
Science Specialist
Austin Independent School District
Austin, Texas

Carol J. Valenta
Senior Vice President
St. Louis Science Center
St. Louis, Missouri

Barry A. Van Deman
President and CEO
Museum of Life and Science
Durham, North Carolina

Senior Editorial Advisors

Napoleon Adebola Bryant, Jr.
Professor Emeritus of Education
Xavier University
Cincinnati, Ohio

Robert M. Jones
Professor of Educational Foundations
University of Houston-Clear Lake
Houston, Texas

Mozell P. Lang
Former Science Consultant
Michigan Department of Education
Science Consultant, Highland Park
   Schools
Highland Park, Michigan

LIFE SCIENCE

Getting Ready for Science — 1

Lesson 1 What Are Tools for Inquiry? — 2
Lesson 2 What Are Inquiry Skills? — 10
Lesson 3 What Is the Scientific Method? — 18
Chapter Review — 24

UNIT A: A World of Living Things

Chapter 1 — Classifying Living Things — 28

Lesson 1 How Are Living Things Classified? — 30
Lesson 2 How Are Plants and Fungi Classified? — 40
Lesson 3 How Are Animals Classified? — 48
Science Projects for Home or School — 59
Chapter Review and Test Preparation — 60

Science Spin Weekly Reader
Technology
Where the Wild Things Are, **56**
People
Bugs Are Cool!, **58**

Chapter 2 — Life Cycles — 62

Lesson 1 What Is Heredity? — 64
Lesson 2 What Are Some Life Cycles of Plants? — 72
Lesson 3 What Are Some Life Cycles of Animals? — 80
Science Projects for Home or School — 91
Chapter Review and Test Preparation — 92

Science Spin Weekly Reader
Technology
Butterfly Tag, **88**
People
Dian Fossey, **90**
Career, **90**

Chapter 3 — Adaptations — 94

Lesson 1 How Do the Bodies of Animals Help Them Meet Their Needs? — 96
Lesson 2 How Do the Behaviors of Animals Help Them Meet Their Needs? — 104
Lesson 3 How Do Living Things of the Past Compare with Those of Today? — 112
Science Projects for Home or School — 123
Chapter Review and Test Preparation — 124

Science Spin Weekly Reader
Technology
Slithering Through the Air, **120**
People
Lydia Villa-Komaroff, **122**
Career, **122**

UNIT B: Looking at Ecosystems

 **Understanding Ecosystems**   **128**

Lesson 1 What Are the Parts of an Ecosystem?   130
Lesson 2 What Factors Influence Ecosystems?   138
Lesson 3 How Do Humans Affect Ecosystems?   148
Science Projects for Home or School   159
Chapter Review and Test Preparation   160

Technology
Aquarius: An Underwater Lab with a View, **156**
People
Meet a Young Conservationist, **158**

 **Energy Transfer in Ecosystems**   **162**

Lesson 1 What Are the Roles of Living Things?   164
Lesson 2 How Do Living Things Get Energy?   172
Science Projects for Home or School   185
Chapter Review and Test Preparation   186

Science Spin Weekly Reader
Technology
On the Prowl, **182**
People
Working with Elephants, **184**
Career, **184**

EARTH SCIENCE

UNIT C: Earth's Changing Surface

Chapter 6 — The Rock Cycle — 190

- Lesson 1 What Are the Types of Rocks? — 192
- Lesson 2 What Is the Rock Cycle? — 200
- Lesson 3 How Do Weathering and Erosion Affect Rocks? — 206
- Lesson 4 What Is Soil? — 214
- **Science Projects** for Home or School — 225
- **Chapter Review and Test Preparation** — 226

Science Spin Weekly Reader

Technology
Crumbling History: Will the Great Sphinx Crumble Like a Cookie?, 222

People
The Color of Dirt, 224
Career, 224

Chapter 7 — Changes to Earth's Surface — 228

- Lesson 1 What Are Some of Earth's Landforms? — 230
- Lesson 2 What Causes Changes to Earth's Landforms? — 238
- Lesson 3 What Are Fossils? — 248
- **Science Projects** for Home or School — 259
- **Chapter Review and Test Preparation** — 260

Science Spin Weekly Reader

Technology
Wipeout! Splash, 256

People
Watching a Volcano, 258

UNIT D: Weather and Space

Chapter 8: The Water Cycle — 264

- Lesson 1 What Is the Water Cycle? 266
- Lesson 2 How Is the Water Cycle Related to Weather? 274
- Lesson 3 How Do Land Features Affect the Water Cycle? 282
- Lesson 4 How Can Weather Be Predicted? 288
- **Science Projects** for Home or School 301
- **Chapter Review and Test Preparation** 302

Science Spin Weekly Reader
Technology Into the Eye of the Storm, 298
People Saving the Earth, 300

Chapter 9: Planets and Other Objects in Space — 304

- Lesson 1 How Do Earth and Its Moon Move? 306
- Lesson 2 How Do Objects Move in the Solar System? 314
- Lesson 3 What Other Objects Can Be Seen in the Sky? 322
- **Science Projects** for Home or School 333
- **Chapter Review and Test Preparation** 334

Science Spin Weekly Reader
Technology Water World, 330
People Moonstruck, 332

PHYSICAL SCIENCE

UNIT E: Matter and Energy

Chapter 10 — Matter and Its Properties — 338

- Lesson 1 How Can Physical Properties Be Used to Identify Matter? — 340
- Lesson 2 How Does Matter Change States? — 348
- Lesson 3 What Are Mixtures and Solutions? — 356
- Science Projects for Home or School — 367
- Chapter Review and Test Preparation — 368

Science Spin Weekly Reader
- **Technology** Fighting Fires with Diapers, 364
- **People** Marie Curie: Scientific Pioneer, 366
- Career, 366

Chapter 11 — Changes in Matter — 370

- Lesson 1 What Is Matter Made Of? — 372
- Lesson 2 What Are Physical Changes in Matter? — 382
- Lesson 3 How Does Matter React Chemically? — 390
- Science Projects for Home or School — 401
- Chapter Review and Test Preparation — 402

Science Spin Weekly Reader
- **Technology** What a Taste Test, 398
- **People** High-flying Scientist, 400
- Career, 400

Chapter 12 — Sound — 404

- Lesson 1 What Is Sound? — 406
- Lesson 2 What Are the Properties of Waves? — 414
- Lesson 3 How Do Sound Waves Travel? — 422
- Science Projects for Home or School — 433
- Chapter Review and Test Preparation — 434

Science Spin Weekly Reader
- **Technology** President's Sunken Ship Discovered, 430
- **People** The Mosquito Kid, 432

Chapter 13 — Light and Heat — 436

- Lesson 1 How Does Light Behave? — 438
- Lesson 2 How Can Heat Be Transferred? — 446
- Lesson 3 How Is Heat Produced and Used? — 454
- Science Projects for Home or School — 465
- Chapter Review and Test Preparation — 466

Science Spin Weekly Reader
- **Technology** Heat from Earth, 462
- **People** A Bright Idea, 464

UNIT F: Forces and Motion

Making and Using Electricity — 470

Lesson 1 What Is Electricity? — 472
Lesson 2 How Are Electricity and Magnetism Related? — 484
Lesson 3 What Are Some Sources of Electricity? — 496
Lesson 4 How Do People Use Energy Resources? — 504
Science Projects for Home or School — 515
Chapter Review and Test Preparation — 516

Science Spin Weekly Reader

Technology
These Boots Were Made for Walking, 512

People
Lighting the Way, 514
Career, 514

Forces and Motion — 518

Lesson 1 How Is Motion Described and Measured? — 520
Lesson 2 What Is Acceleration? — 528
Lesson 3 Why Is the Force of Gravity Important? — 536
Science Projects for Home or School — 547
Chapter Review and Test Preparation — 548

Science Spin Weekly Reader

Technology
Science Soars at the Olympics, 544

People
Testing the Wind, 546

Simple Machines — 550

Lesson 1 How Do Simple Machines Help People Do Work? — 552
Lesson 2 How Do a Pulley and a Wheel-and-Axle Help People Do Work? — 560
Lesson 3 How Do Other Simple Machines Help People Do Work? — 568
Science Projects for Home or School — 579
Chapter Review and Test Preparation — 580

Science Spin Weekly Reader

Technology
Gadget Guy, 576

People
Flying High, 578
Career, 578

References — 582

Health Handbook — R1
Reading in Science Handbook — R16
Math in Science Handbook — R28
Safety in Science — R36
Glossary — R37
Index — R46

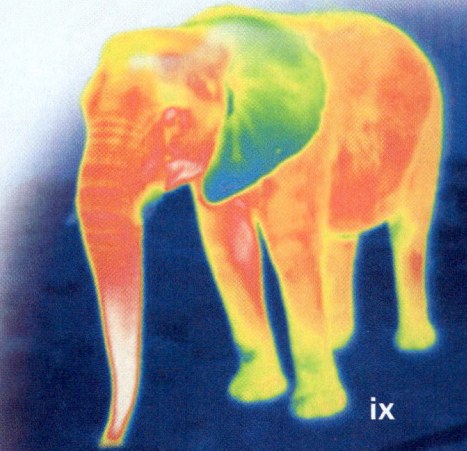

UNIT B
Looking at Ecosystems

| Chapter 4 | Understanding Ecosystems |
| Chapter 5 | Energy Transfer in Ecosystems |

LIFE SCIENCE

Coyne Center Elementary Theme Gardens

TO: maria@hspscience.com
FROM: steven@hspscience.com
RE: Milan, Illinois

Dear Maria,
You asked me where I learned how to be a gardener. Believe it or not, I learned at school! My school has a schoolyard habitat that has 12 outdoor garden areas. We even have a pizza garden! We grow all of the toppings to make a delicious pizza!
Your pen pal friend,
Steven

Colorado's Ocean Journey

TO: jalen@hspscience.com
FROM: destiny@hspscience.com
RE: Colorado

Dear Jalen,
At Colorado's Ocean Journey, you can see a model of the long route a fish has to swim to get from the Colorado River to the sea. Lots of plants and animals rely on the water to survive. You get to see those up close as you travel along the way. That is quite a journey!
Your pen pal,
Destiny

Experiment!

Counting Species When people develop land and build new houses, the environment that was already there is changed. The animals living in the habitat must find new places to live. Is there a difference between natural habitats and habitats that humans have developed? Are there more types of living things in one of the two kinds of environments? Plan and conduct an experiment to find out.

Chapter 4
Understanding Ecosystems

Lesson 1 What Are the Parts of an Ecosystem?

Lesson 2 What Factors Influence Ecosystems?

Lesson 3 How Do Humans Affect Ecosystems?

Vocabulary
environment
ecosystem
population
community
biotic
abiotic
diversity
pollution
habitat restoration

What do YOU wonder?

This hawk has very sharp claws, called talons. Why don't birds such as sparrows have talons?

Lesson 1

What Are the Parts of an Ecosystem?

Fast Fact

Silver Kings The tarpon in this photograph are not yet full-grown! These fish don't become adults until they are between 7 and 13 years old, when they can weigh more than 91 kilograms (200 lb). Tarpon live in salt water, but they can survive in a variety of ecosystems. In the Investigate, you will observe how sunlight affects plants in another ecosystem.

Investigate

Modeling an Ecosystem

Materials
- gravel
- 6 small plants
- 2 empty 2-L soda bottles with tops cut off
- sand
- water in a spray bottle
- soil
- clear plastic wrap
- 2 rubber bands

Procedure

1. Pour a layer of gravel, a layer of sand, and then a layer of soil into the bottom of each bottle.

2. Plant three plants in each bottle.

3. Spray the plants and the soil with water. Cover the top of each bottle with plastic wrap. If necessary, hold the wrap in place with a rubber band.

4. Put one of the terrariums you just made in a sunny spot. Put the other one in a dark closet or cabinet.

5. After three days, observe each terrarium and record what you see.

Draw Conclusions

1. What did you observe about each of your ecosystems after three days? What part was missing from one ecosystem?

2. **Inquiry Skill** Scientists often learn more about how things affect one another by making a model. What did you learn by making a model and observing how its parts interact?

Step 2

Step 3

Investigate Further

What effect does sunlight have on seeds that have just been planted? First, write your hypothesis. Then plan an experiment to see if your hypothesis is supported.

Reading in Science

VOCABULARY
environment p. 132
ecosystem p. 132
population p. 134
community p. 136

SCIENCE CONCEPTS
▶ how living and nonliving parts of an ecosystem interact
▶ what populations and communities are

READING FOCUS SKILL
MAIN IDEA AND DETAILS
Look for the parts that make up an ecosystem.

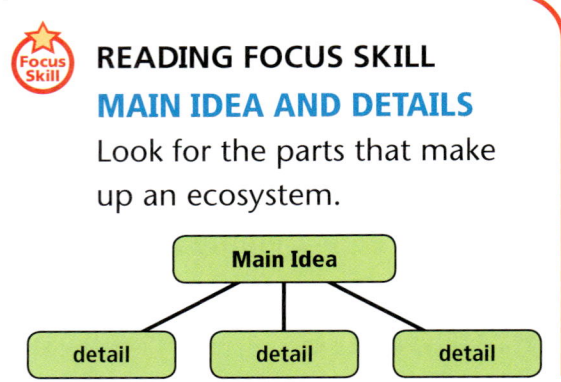

Ecosystems

Where do you live? You might name your street and town. You also live in an environment. An **environment** is all the living and nonliving things that surround you. The living things in your environment are people, other animals, and plants. The nonliving things around you include water, air, soil, and weather.

The parts of an environment affect one another in many ways. For example, animals eat plants. The soil affects which plants can live in a place. Clean air and clean water help keep both plants and animals healthy. All the living and nonliving things in an area form an **ecosystem** (EE•koh•sis•tuhm).

An ecosystem can be very small. It might be the space under a rock. That space might be home to insects and tiny plants. You might need a microscope to see some of the things living there.

This prairie smoke plant grows well in the hot, dry climate of prairies and grasslands. ▼

Prairie dogs also live on the prairies and grasslands. ▼

Moose thrive in a coniferous, or evergreen, forest ecosystem.

The small ecosystem found under a rock has nonliving parts, too. They include pockets of air and the soil under the rock. You might find a few drops of water or maybe just damp soil. All ecosystems must have at least a little water.

The ecosystem under this rock has a climate. The *climate* in an area is the average weather over many years. Climate includes temperature and rainfall. The climate of an ecosystem depends on where the ecosystem is. If this rock is in Florida, its climate is warm and wet. If the rock is in Maine, its winters are icy.

An ecosystem can also be as large as a forest. A forest can provide many kinds of food and shelter. This ecosystem may include hundreds of kinds of plants and animals. Each organism finds what it needs in the forest.

Like all ecosystems, a forest has nonliving parts. They include water, air, soil, and climate. Later, you will read more about ways living and nonliving parts of an ecosystem affect one another.

 MAIN IDEA AND DETAILS Name the two parts of an ecosystem, and give two examples of each part.

This individual waterlily is part of a large population of waterlilies.

Individuals and Populations

One plant or animal is an *individual*. For example, one blueberry bush is an individual. One honeybee is an individual. One blue jay is an individual. You are an individual.

A group made up of the same kind of individuals living in the same ecosystem is a **population**. A group of blueberry bushes is a population. So is a hive of bees. So are all the blue jays living in one forest. So are all the people living in one city.

Robins might live in the same forest as the blue jays. Robins are a different kind of bird. That makes them a different population.

The members of a population might not live in a group. For example, frogs don't live in families. Still, a number of green tree frogs may live near the same pond. They belong to the same population. Bullfrogs might also live near that pond. They are a different population.

Many animals live in groups. People live in families. How many people are in your family? Wolves live in packs. A pack can have from 3 to 20 wolves. A wolf population may have several packs. The wolf population in Yellowstone National Park includes 19 packs.

134

Some populations can live in more than one kind of ecosystem. For instance, red-winged blackbirds often live in wetlands, but they are also found in other areas. Red-winged blackbirds can live in different ecosystems. If one ecosystem no longer meets the needs of these birds, they fly to another one.

Some populations can live in only one kind of ecosystem. One such animal is the Hine's emerald dragonfly. This insect can live only in certain wetlands. It can't survive in other places. Because this dragonfly can live only in specific places, its total number is very small.

Ecosystems are often named for the main population that lives there. For example, one kind of ecosystem forms where a river flows into the ocean. There, fresh water mixes with salt water. Many trees can't live in salty water. But mangrove trees have roots that allow them to get rid of the salt in the water. When many mangrove trees live in a salty ecosystem, the area is called a *mangrove swamp*.

 MAIN IDEA AND DETAILS Name an individual and a population that are not mentioned on these two pages.

Eeek! Oh System!

Work with a partner to list some of the populations in your school ecosystem. Think about the building and the land around it. Then compare lists with other students. Did you list the same populations?

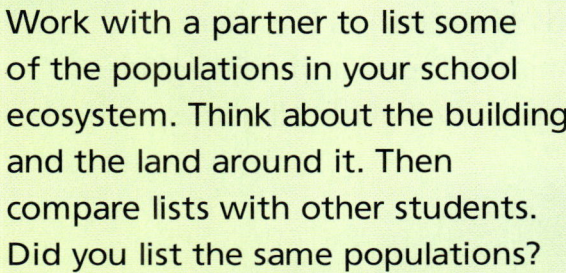

This individual male red-winged blackbird is part of a large population of blackbirds.

A population of red-winged blackbirds can include several million birds. Some of the birds fly 80 kilometers (50 mi) to find food.

Communities

You live in a community. Other animals and plants do, too. A **community** is all the populations that live in the same place.

Have you visited the Everglades National Park? Many different populations make up this community. The plants include mangrove trees, cypress trees, and saw grass. If you have been to the Everglades, you may know about the mosquitoes from getting bitten! The area has 43 kinds. And 50 kinds of butterflies live there.

Animals found in the Everglades community include alligators, bobcats, and raccoons. Bird-watchers like to visit the Everglades. They try to see some of the 350 kinds of land birds and 16 kinds of wading birds that live there.

In some ways, the Everglades is like all communities. The plants and animals there depend on one another. Some animals eat the plants. Other animals eat the plant eaters. The animals help spread the plants' seeds. The plants provide shelter for the animals.

 MAIN IDEA AND DETAILS Name three populations that might be found in a forest community.

▼ Many populations make up the communities in this cold taiga ecosystem. They include evergreen trees, moose, and many kinds of birds.

Reading Review

 1. MAIN IDEA AND DETAILS Draw and complete this graphic organizer.

A pond ecosystem is made up of living and nonliving things.

Living Things
A _____

Nonliving Things
B _____

2. SUMMARIZE Write a summary of this lesson by using the lesson vocabulary words in a paragraph.

3. DRAW CONCLUSIONS Why do some ecosystems include more living things than other ecosystems?

4. VOCABULARY Use the lesson vocabulary words to create a matching quiz.

Test Prep

5. Critical Thinking How is a population different from a community?

6. Which word describes a group of cows standing together?
- **A.** community
- **B.** ecosystem
- **C.** individual
- **D.** population

Links

Writing

Expository Writing
You are a scientist planning an ecosystem on the moon. Write **two paragraphs** explaining what this ecosystem should include.

Math

Solve a Problem
The Everglades includes many "rivers of grass." The water in these rivers moves slowly, only 30 meters (100 ft) a day. How many meters would the water move in June? In February?

Social Studies

Ecosystems and People
Choose a group of people who live in an ecosystem different from yours. Find out how that ecosystem affects the people. Share what you learn in an oral or written report.

 **For more links and activities**, go to www.hspscience.com

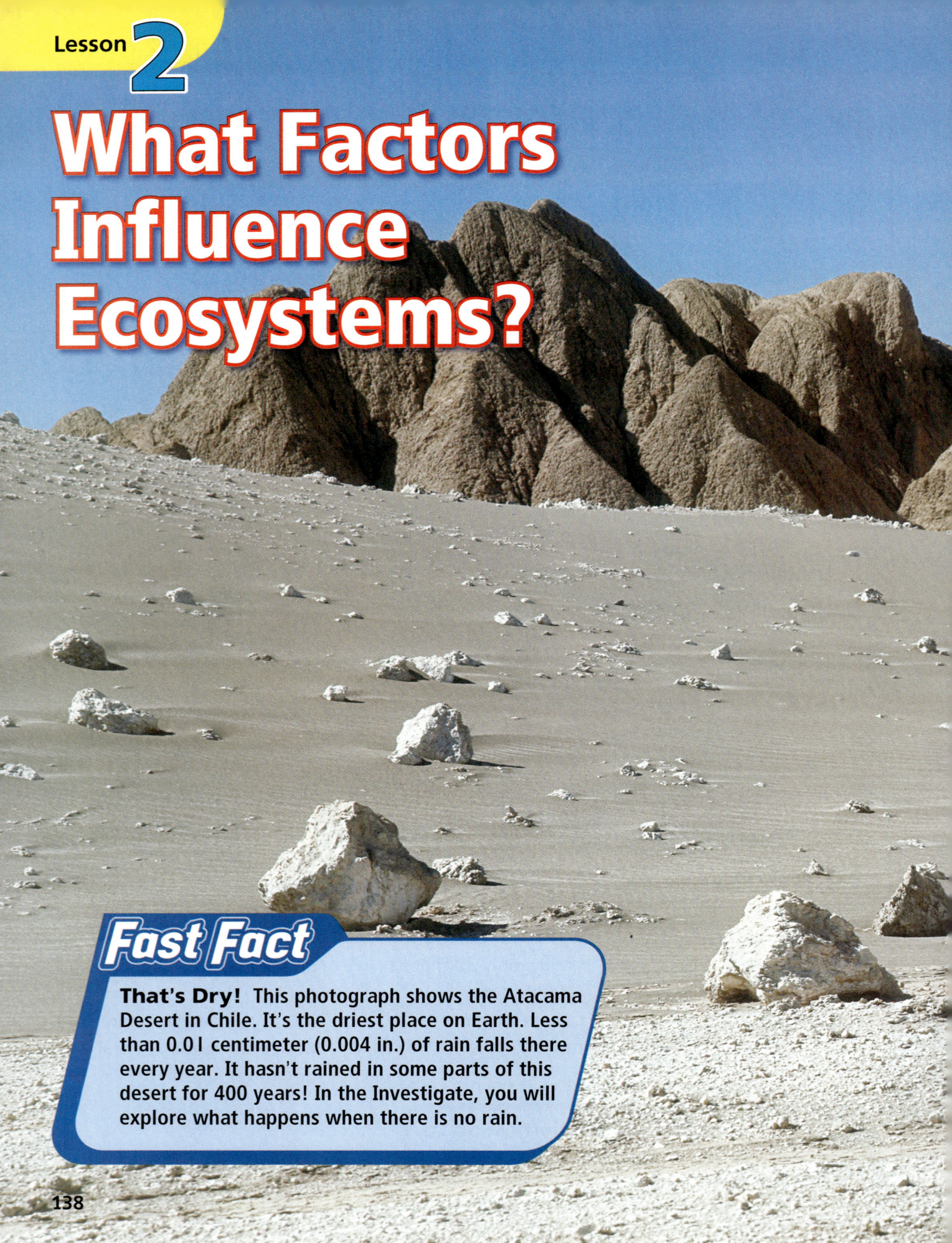

Lesson 2

What Factors Influence Ecosystems?

Fast Fact

That's Dry! This photograph shows the Atacama Desert in Chile. It's the driest place on Earth. Less than 0.01 centimeter (0.004 in.) of rain falls there every year. It hasn't rained in some parts of this desert for 400 years! In the Investigate, you will explore what happens when there is no rain.

Investigate

Observing the Effects of Water

Materials
- 4 small identical plants in clay pots
- water
- large labels

Procedure

1. Use the labels to number the pots 1, 2, 3, and 4. Label pots 1 and 2 *watered*. Label pots 3 and 4 *not watered*.

2. Make a table like the one shown here. Draw a picture of each plant under Day 1.

	Day 1	Day 4	Day 7	Day 10
Plant 1 (watered)				
Plant 2 (watered)				
Plant 3 (not watered)				
Plant 4 (not watered)				

3. Place all four pots in a sunny window.

4. Water all four pots until the soil is a little moist. Keep the soil of pots 1 and 2 moist during the whole experiment. Don't water pots 3 and 4 again.

5. Wait three days. Then observe and record how each plant looks. Draw a picture of each one under Day 4.

6. Repeat Step 5 twice. Draw pictures of the plants on Days 7 and 10.

Draw Conclusions

1. What changes did you observe during this Investigate? What do they tell you?

2. **Inquiry Skill** Scientists compare changes to determine how one thing affects another. How could you compare how fast the soil dries out in a clay pot with how fast it dries out in a plastic pot?

Investigate Further

How does covering a plant with plastic wrap affect the plant's need for water? Write your hypothesis. Then design and carry out an experiment to check your hypothesis.

Reading in Science

VOCABULARY
biotic p. 140
abiotic p. 142
diversity p. 146

SCIENCE CONCEPTS
▶ how biotic and abiotic factors affect ecosystems
▶ how climate influences an ecosystem

 READING FOCUS SKILL
CAUSE AND EFFECT
Look for ways in which factors affect ecosystems.

Living Things Affect Ecosystems

Do plants and animals need each other? Yes, they do! Plants and animals are living parts of an ecosystem. These living parts are **biotic** factors. *Bio* means "life." Biotic factors affect the ecosystem and one another in many ways.

For example, plants provide food for caterpillars, birds, sheep, and other animals. People eat plants every day—at least they should.

Plants also provide shelter for animals. For instance, many insects live in grasses. Squirrels make dens in trees. Your home likely contains wood from trees.

Animals help plants, too. When animals eat one kind of plant, it can't spread and take over all the available space. This gives other kinds of plants room to grow.

A gypsy moth can lay 1000 eggs or more. Most of the eggs hatch into hungry caterpillars like this one. ▶

A healthy tree isn't hurt when a few insects nibble on it.

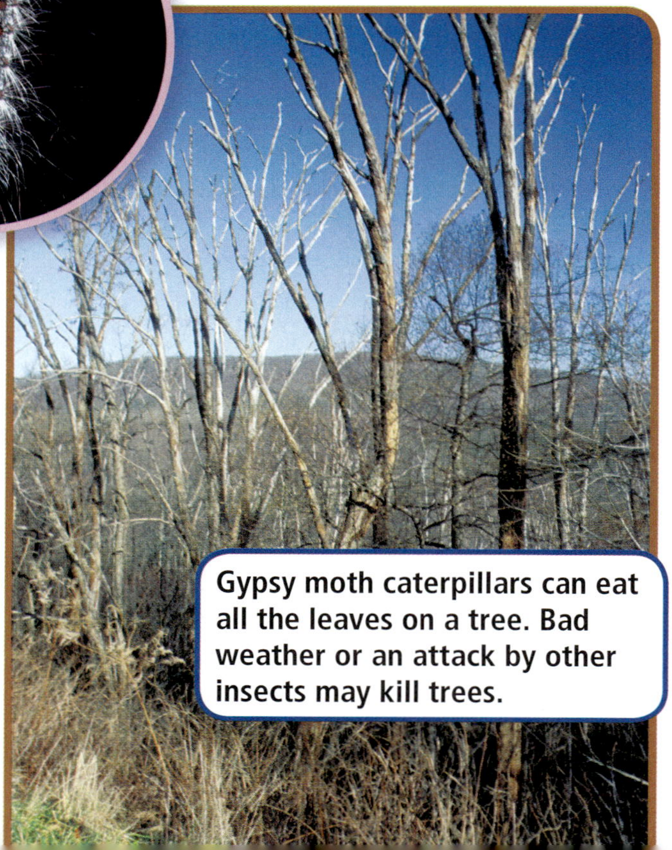

Gypsy moth caterpillars can eat all the leaves on a tree. Bad weather or an attack by other insects may kill trees.

Animals help plants in other ways. Animal droppings make the soil richer. Earthworms help loosen the soil. Rich, loose soil helps plants grow.

At the same time, too many plant eaters can be harmful. A herd of hungry deer can eat enough leaves to kill a tree. A huge swarm of locusts can leave a field bare of plants.

You know that animals affect one another. For example, wolves eat rabbits. If the wolf population becomes too large, wolves can wipe out the rabbits. Then the wolves go hungry. Without the rabbits to eat them, the grasses spread.

In this case, an increase in wolves causes a decrease in rabbits. Fewer rabbits causes an increase in plants.

A change in plants can also cause a change in animals. If dry weather or disease kills the grasses, the rabbits starve. Then the wolves go hungry, too. Disease can also kill animals in an ecosystem.

Sometimes, a new kind of plant or animal changes an ecosystem. For example, people brought the skunk vine to the United States from Asia in 1897. For a time, they planted it as a crop. Now it grows wild. This smelly vine can grow 9 meters (30 ft) long! It crowds out other plants, and it can even grow underwater.

CAUSE AND EFFECT
Explain how an increase in plants could affect an ecosystem.

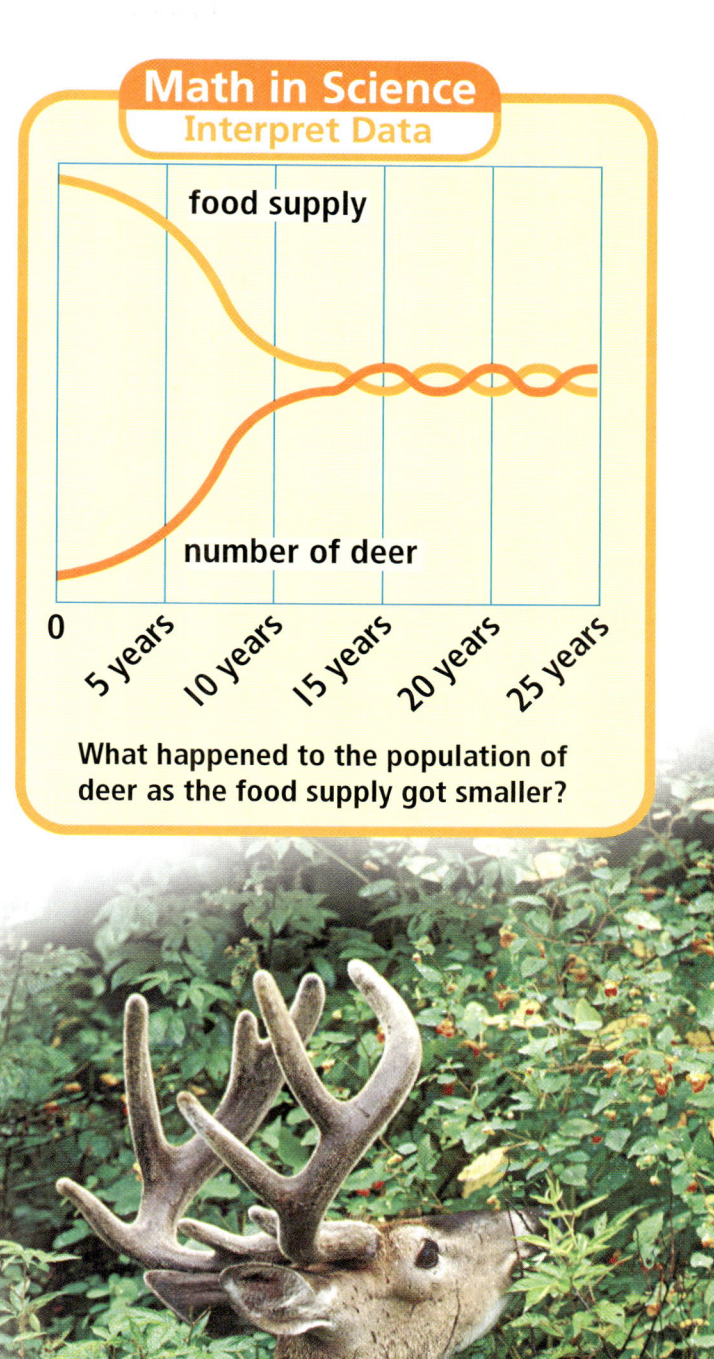

Math in Science
Interpret Data

What happened to the population of deer as the food supply got smaller?

Tree leaves are a main source of food for deer. It takes 15 to 30 acres of land to provide enough food for one deer.

141

Nonliving Things Affect Ecosystems

Plants and animals are the living parts of an ecosystem. The nonliving parts include sunlight, air, water, and soil. The nonliving parts are **abiotic** factors. They are just as important as the biotic factors.

For example, a change in the water supply can affect all the living things in an ecosystem. Too little rain causes many plants to wilt and die. Animals must find other homes. Some may die.

An ecosystem with rich soil has many plants. Where the soil is poor, few plants grow. Few plants mean few animals in the ecosystem.

Air, water, and soil can contain harmful substances. They can affect all living things. You will learn more about this problem later in the chapter.

 CAUSE AND EFFECT How might a change in the water supply affect a rabbit?

Science Up Close

Nonliving Factors

Without the nonliving parts of an ecosystem, there would be no living parts.

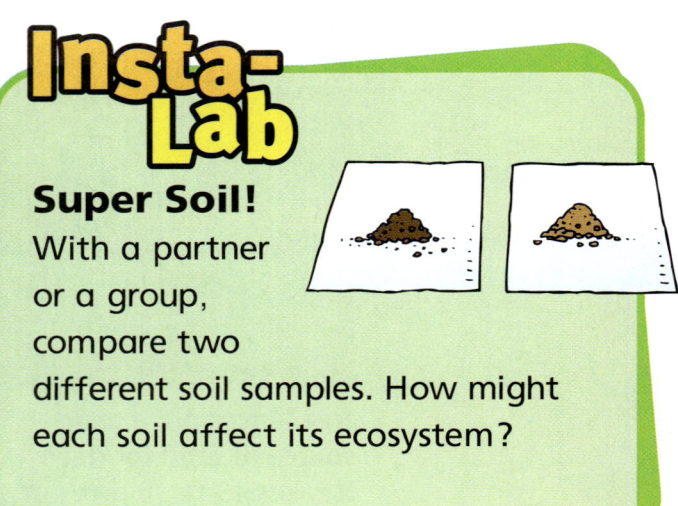

Insta-Lab

Super Soil! With a partner or a group, compare two different soil samples. How might each soil affect its ecosystem?

Climate Affects Ecosystems

What is the climate like where you live? Is it warm and sunny, or is it cool and rainy? Maybe it's something in between.

Climate is an abiotic factor. It's a combination of other abiotic factors. Climate includes the amount of rainfall and sunlight in a region. It also includes the repeating patterns of the temperature of the air during the year.

Climate affects the soil. Some climates allow many plants to grow and help dead plants decay. Animals that eat the plants leave behind their droppings. The decaying plants and droppings make the soil richer.

Climate affects the kinds of plants and animals in an ecosystem. For example, warm, wet climates support tropical rain forests. Hot summers and cold winters result in temperate forests.

The frozen tundra suits the hardy caribou. The mosses they eat thrive there. Zebras could not survive in the tundra. They need the mild climate and tender grasses of the savanna.

 CAUSE AND EFFECT What would happen to an ecosystem if its climate changed?

World Climate Zones

This map shows six climate zones around the world.

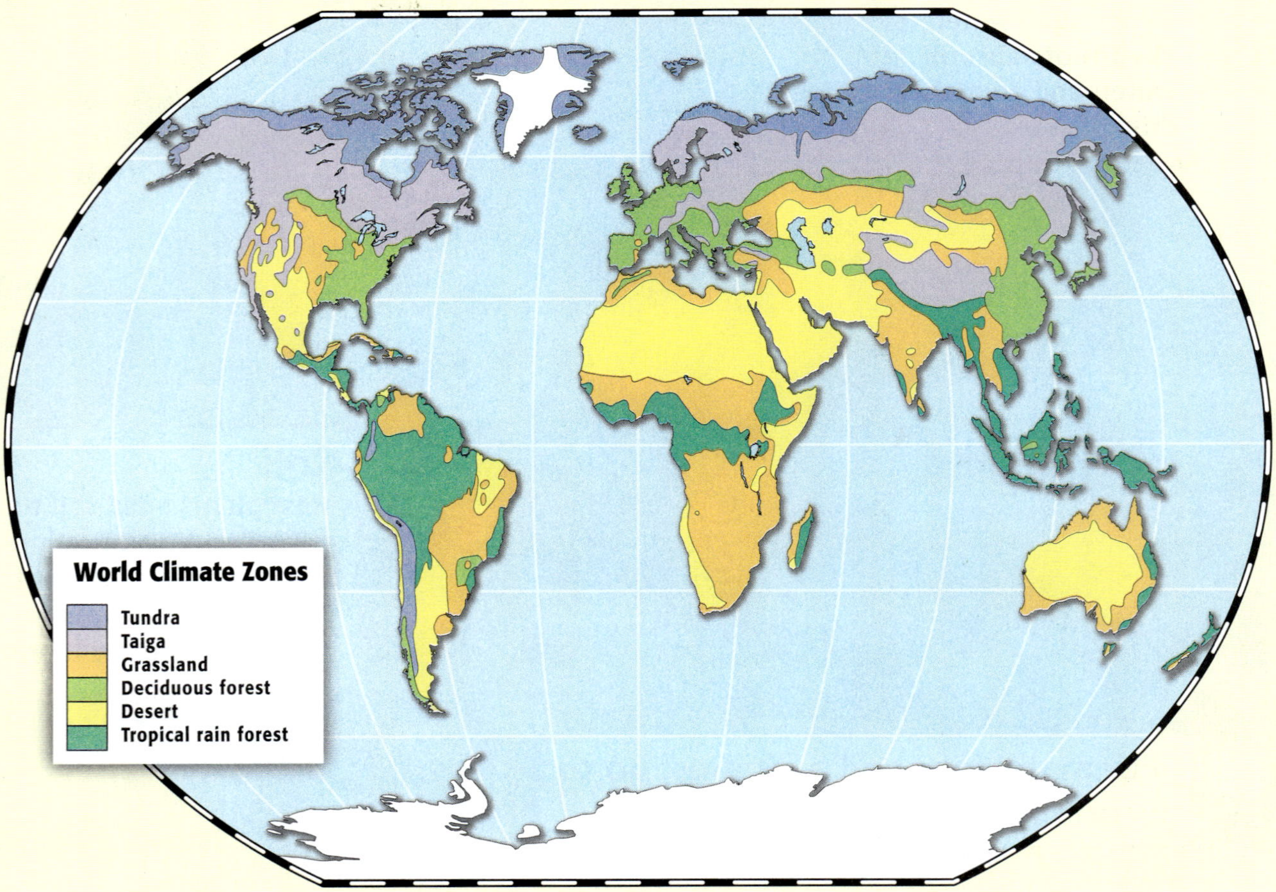

World Climate Zones
- Tundra
- Taiga
- Grassland
- Deciduous forest
- Desert
- Tropical rain forest

Deciduous forests have four seasons. The trees, such as oaks and maples, lose their leaves in the fall. This helps them survive the cold winters.

Rain forests receive 2000 to 10,000 millimeters (7 to 33 ft) of rain each year! Tropical rain forests are near the equator.

The climate in the grassy savanna is nearly the same all year. The temperature stays between 18°C and 22°C (64°F and 72°F).

Deserts get only about 250 millimeters (10 in.) of rain a year. Plants there grow very quickly after a rain. Their seeds can survive for years as they wait for more rain.

The taiga covers more of Earth than any other kind of plant community. The taiga is mostly just south of the tundra and is very cold in winter. Most of its trees are evergreens.

The tundra has the coldest climate: −40°C to 18°C (−40°F to 64°F). *Tundra* means "treeless plain."

There are layers in a rain forest.

The *canopy* is the upper part of the trees. It is home to most rain-forest animals.

The next layer is the cool, dark *understory.* This layer is just right for plants that grow well in shade.

The bottom layer is the *forest floor,* where decaying matter provides food for plants.

Diversity

A rain-forest ecosystem provides many sources of food and shelter. That's why it has the most diversity of all of Earth's ecosystems. **Diversity** refers to the number of different kinds of living things.

In a rain forest, a wide range of plants and animals can find what they need to survive. Many kinds of monkeys live in the treetops. Snakes slip from branch to branch. Bright butterflies flit among the flowers. Frogs of many colors cling to tree trunks. Mushrooms and earthworms hide under decaying leaves. Some rain-forest plants have giant leaves. Other plants can't be seen without using a microscope.

Some ecosystems don't have much diversity. The tundra, for example, is very cold and dry. Much of its soil is frozen. Few living things can survive there.

How much diversity does the ecosystem where you live have?

CAUSE AND EFFECT What leads to a diversity of living things in an ecosystem?

Reading Review

1. **CAUSE AND EFFECT** Draw and complete the graphic organizer.

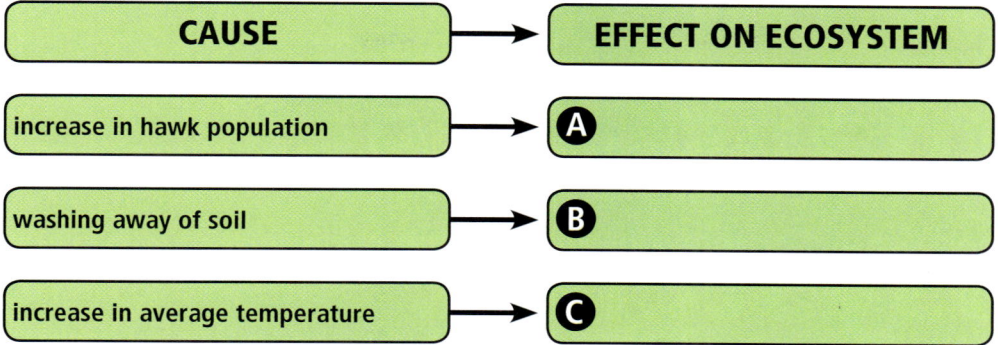

2. **SUMMARIZE** Use your completed graphic organizer to write a lesson summary.

3. **DRAW CONCLUSIONS** Which can exist without the other—biotic factors or abiotic factors? Explain your answer.

4. **VOCABULARY** Write a quiz-show-type question for each of the vocabulary words.

Test Prep

5. **Critical Thinking** How might flooding in their ecosystem affect some robins?

6. Which of these is an abiotic factor in an ecosystem?
   A. ant   C. earthworm
   B. decaying plant D. sand

Links

Writing
Persuasive Writing
Write a **travel brochure** for a climate zone where few people vacation, such as the tundra or taiga. Tell your readers what interesting things they can see and experience there.

Math
Make a Graph
Find the average rainfall in five of the six world climate zones, including your own region. Then make a bar graph that compares the rainfalls.

Literature
Learn More
Read a current nonfiction book about one of the world climate zones, such as the desert. After learning more about that climate zone, share what you know by making a display or a written report.

For more links and activities, go to www.hspscience.com

Lesson 3

How Do Humans Affect Ecosystems?

Fast Fact

Saving Soil Contour-plowing slopes helps keep rain from washing away the soil. It can reduce erosion by as much as 75 percent! In the Investigate, you will experiment with "rain" and erosion.

Investigate

Losing It: Observing Erosion

Materials
- marker
- soil and small rocks
- 2 paper cups
- water
- 2 clean plastic foam trays
- sharpened pencil
- measuring cup

Procedure

1. Use the marker to write *A* on one tray and on one paper cup. Write *B* on the second tray and on the second paper cup.

2. In each tray, make an identical slope out of soil and rocks.

3. Carefully use the pencil to make three small holes in the bottom of cup A. Make six larger holes in the bottom of cup B.

4. **Record** how the two slopes look now. Label your drawings *A* and *B*.

5. Hold cup A over the slope in tray A. Slowly pour 1 cup of water into the paper cup, and let it run down the slope. **Record** how the slope looks now.

6. Repeat Step 5, using cup B and tray B. Then **record** how the slope looks.

Step 2

Step 5

Draw Conclusions

1. At the end of the activity, compare the slopes. What did each cup represent?

2. **Inquiry Skill** An **experiment** is a careful, controlled test. What were you testing and what did you control in the activity?

Investigate Further

Try the same activity, using only rocks, using level soil, or using plants growing in the soil. Make a **prediction** about what will happen, and then do the activity to see if your **prediction** was accurate.

149

Reading in Science

VOCABULARY
pollution p. 152
habitat restoration p. 153

SCIENCE CONCEPTS
▶ how humans use the resources in ecosystems
▶ the positive and negative ways humans affect ecosystems

READING FOCUS SKILL
COMPARE AND CONTRAST
Compare positive and negative effects that humans have on ecosystems.

| alike | different |

Humans Within Ecosystems

Do you use any natural resources? You do if you breathe! Natural resources are the parts of ecosystems that humans use, including air.

What other natural resources do you use? Do you ever go to the seashore or a park? Those are natural resources. When you turn on a light, you use natural resources. Most electricity is produced by burning coal. Coal is a natural resource that is taken from under the ground.

Do you ride a bus to school? The fuel that makes the bus run is made from oil. Oil is a natural resource that is also taken from under the ground.

Minerals are natural resources, too. Iron, copper, and aluminum are examples of mineral resources.

Some natural resources can be replaced. Sunlight, air, and water are renewable resources. People can grow more trees and plant more crops.

▼ Natural resources include lakes, fresh air, and sunlight.

Fish and other living things are also natural resources.

150

More than 2000 years ago, people drank tea made from the bark of the white willow tree to help ease pain. In 1829, scientists discovered the chemical in the willow that reduces pain. They used it to make aspirin tablets.

Aspirin can be bought without a prescription. However, it is a powerful and valuable drug.

People have been growing wheat for 10,000 years.

Making bread is just one way wheat is used. This grain is also used in cakes, cookies, cereals, and pastas. Parts of the wheat plant are fed to cattle.

Some natural resources can't be replaced. They include coal, gas, and oil. After the supplies buried underground are used, these resources will be gone.

Humans use natural resources in many ways. People build homes and furniture from wood. They make bricks from clay, and glass from sand. They use iron to make steel, which they then use to make cars and many other things.

People raise crops to feed themselves and their animals. They use plants as medicines, too. Humans learned long ago that plants could help treat or cure some illnesses. More than 40 percent of the medicines used today originally came from plants.

For example, a medicine made from a plant called foxglove can help treat heart disease. Scientists use the bark of the Pacific yew tree to make a medicine to treat cancer.

Scientists have tested only 2 percent of all plants to see if they can be used as medicines. Who knows how many more medicines plants may provide?

 COMPARE AND CONTRAST Compare the supply of crops with the supply of coal.

151

Negative and Positive Changes

Humans make many negative changes in ecosystems. When people clear land for houses and shopping malls, they destroy habitats. As a result, the animals that lived there can no longer meet all their basic needs. They must move or die.

Farmers plow land to plant crops. Plowing loosens soil. That makes it easier for rain and wind to carry away the soil. Humans also cause some kinds of pollution. **Pollution** happens when harmful substances mix with water, air, or soil.

Storms washing chemicals off fields can cause water pollution. These chemicals flow into streams and rivers. Trash and waste from homes and businesses can also enter the water supply.

Much air pollution comes from burning gasoline. Fumes from car engines carry chemicals into the air. Factory smokestacks release more chemicals. Some of these chemicals form acid rain. Acid rain can burn trees and other plants. It can poison lakes and rivers.

Soil pollution can come from fertilizers and trash. Wastes, such as old paint and drain cleaners, can poison the soil.

Water treatment plants remove harmful substances from water before it reaches people's homes. ▶

Cars and other vehicles are a major source of air pollution in cities. This pollution causes smog and breathing problems.

Bicycles don't release pollution. They also provide a good way to get exercise.

Many laws are designed to prevent water pollution, but it still happens.

152

Without plant roots to hold soil in place, much of the soil can wash away.

These people are helping prevent beach erosion by planting dune grasses.

Humans also make positive changes. Many groups are working to repair damage to ecosystems. They plant new trees and create new wetlands. They build parks over closed landfills. This process is **habitat restoration**.

People are also polluting less. For example, cars now have special devices on their tailpipes. These devices reduce the harmful gases that escape into the air. Factories now release fewer chemicals. They don't dump wastes into rivers and streams.

Many people now use natural ways to get rid of weeds and insects. They spread fewer chemicals on fields and lawns.

People also recycle paper, glass, metal, and plastic. Recycling uses less energy than making new products. That means less coal is burned. Burning less coal means less pollution.

People are learning other ways to help reduce pollution. Science is one way of finding solutions to the problems caused by pollution.

 COMPARE AND CONTRAST Which kind of pollution is most harmful—water, air, or soil? Why?

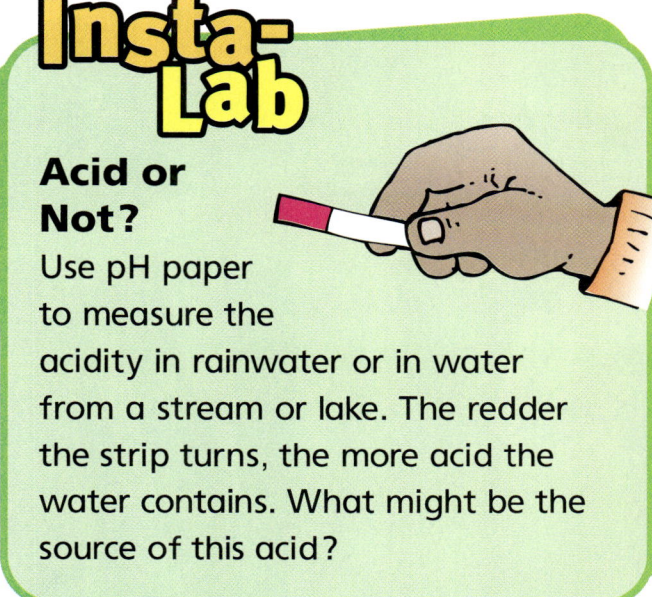

Acid or Not?

Use pH paper to measure the acidity in rainwater or in water from a stream or lake. The redder the strip turns, the more acid the water contains. What might be the source of this acid?

Planning for Change

Earth's human population keeps growing. People need more space for places to work and live. But before we build, we have to consider both abiotic and biotic factors. Some abiotic factors include the type of soil, the amount of rain, and the climate. A building set on soft soil will not stand. A home must be able to withstand the weather of the area where it is built.

Biotic factors can affect more than humans. Building new structures often means destroying wildlife habitats. Builders should plan new projects in ways that protect ecosystems. A building near a river must not pollute the water. Even working close to a river can cause problems. Soil can wash into the water. Too much soil in the water can harm fish and plants.

Wetlands near the river must not be filled in. Wetlands help keep the water clean and provide homes for many plants and animals. Builders often leave or create ponds or pockets of forest. These habitats provide homes for some wildlife. Every ecosystem has a delicate balance. People must do their part to protect that balance.

 COMPARE AND CONTRAST Compare a human and an animal seeking a new habitat. How are they the same?

Construction must be carefully planned to protect natural ecosystems. ▶

◀ Land-use planners must consider the biotic and abiotic factors in an ecosystem.

Reading Review

 1. COMPARE AND CONTRAST Draw and complete this graphic organizer.

```
            Human Effects on Ecosystems
   ┌─────────────────────────┐   ┌─────────────────────────┐
   │ Negative Change:         │   │ Positive Change: A       │
   │ destruction of habitats  │   │                          │
   └─────────────────────────┘   └─────────────────────────┘
   ┌─────────────────────────┐   ┌─────────────────────────┐
   │ Negative Change: B       │   │ Positive Change: devices │
   │                          │   │ to reduce air pollution  │
   │                          │   │ from cars                │
   └─────────────────────────┘   └─────────────────────────┘
```

2. SUMMARIZE Write two sentences that tell what this lesson is mainly about.

3. DRAW CONCLUSIONS Why do humans affect natural ecosystems in negative ways?

4. VOCABULARY Make an acrostic for *pollution*. Each letter begins a sentence about that term.

Test Prep

5. Critical Thinking How can a weedkiller used on a cornfield pollute a lake miles away?

6. Which of these is soil pollution likely to cause?
   A. air pollution   **C.** abiotic factors
   B. biotic factors   **D.** water pollution

Links

Writing
Persuasive Writing
Think of a way your community has had a negative effect on an ecosystem. Write a **letter** you might send to the editor of a local newspaper describing the problem.

Math
Comparing Gas Use
An older, larger car gets 10 miles per gallon of gasoline. A newer, smaller car gets 35 miles per gallon. How many fewer gallons would the newer car use on a 70-mile trip?

Health
Pollution and You
With a small group, research the health effects of water, air, or soil pollution. Then share what you learn in a written report, an oral presentation, or a poster.

 **For more links and activities**, go to www.hspscience.com

AQUARIUS
An Underwater Lab with a View

Many U.S. kids share their bedrooms with a brother or sister. Sharing a room can be a pain. Six scientists know it! They squeezed into a small underwater laboratory to study a coral reef. They worked and slept in the tiny space for ten days.

"I've been on missions where people snored," said Celia Smith, who led the mission in October. "You just kind of kick their bunk and try to get to sleep before they do."

This buoy supplies the *Aquarius* with fresh air and electricity.

▲ An aquanaut peers through a window in the *Aquarius* as another aquanaut swims around the laboratory.

The underwater lab is called *Aquarius*. It looks like a little yellow submarine.

The *Aquarius* was placed near the Florida Keys National Marine Sanctuary, a protected area of the ocean. Each year, several teams of scientists have lived in *Aquarius* for up to ten days at a time. The scientists who live and work in *Aquarius* are called aquanauts (AH•kwuh•nawts). The aquanauts study the nearby coral reef and the creatures that live in it.

A Room with a View

The latest team to visit *Aquarius* says the best part of living there is the view. They can see colorful fish swimming in the nearby reef. Smith said it's hard to tell, though, whether the aquanauts are watching the fish or the fish are watching the aquanauts.

The aquanauts spend as much time as they can on the reef studying sea life.

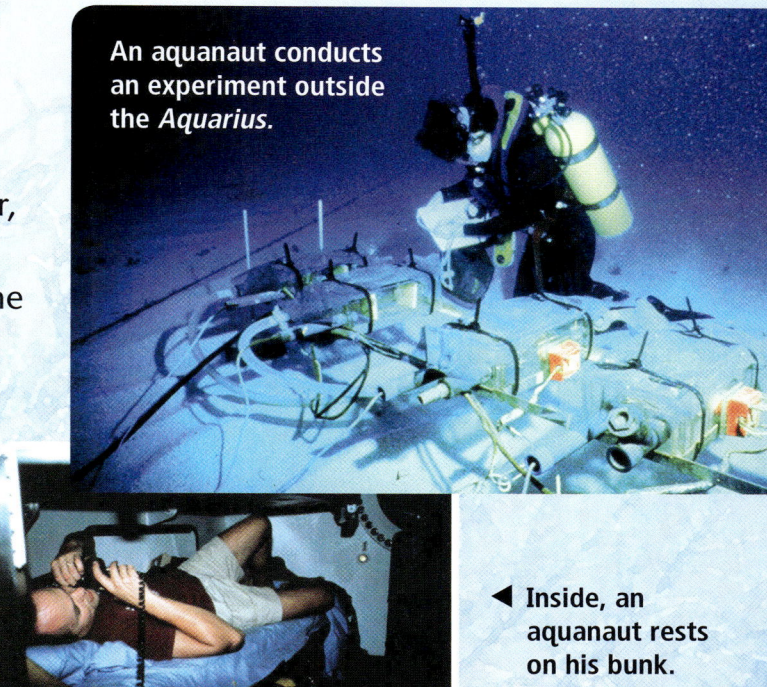

An aquanaut conducts an experiment outside the *Aquarius*.

◀ Inside, an aquanaut rests on his bunk. He doesn't have much room.

They can spend up to nine hours at one time outside the laboratory.

Smith likes to remind people that humans have barely begun to explore the oceans and need to learn more about life in the deep. "The really important thing for us to realize is how much we don't know about the oceans," she said.

THINK ABOUT IT

1. How might living underwater help scientists learn more about a coral reef ecosystem?
2. What might be the best thing about living underwater? What might be the worst thing?

IN THE DEEP

Aquarius is located about 19 meters (63 ft) deep in the ocean. It has sleeping space for six, a bathroom, a trash compactor, and computer stations.

Electrical cables and tubes connect the *Aquarius* to a buoy on the ocean's surface. The tubes carry fresh air to the *Aquarius*, and the cables supply electricity.

The work the aquanauts are doing is expected to help NASA. NASA scientists say that living and working on the *Aquarius* is similar to what it will be like to live and work in a space station. They hope to better prepare astronauts for space by studying how the aquanauts live and work underwater.

Find out more! Log on to www.hspscience.com

SCIENCE Spin from WEEKLY READER

People

Meet a Young CONSERVATIONIST

Fourth grader Blake Wichtowski told people at last year's Kids' Summit that wild blue lupine flowers would help the endangered Karner blue butterfly. Officials from New York are turning this idea into a reality.

Blue lupine is the only food that Karner caterpillars will eat.

With the help of the Seneca Park Zoo in Rochester and other officials, Blake's fourth-grade class and a class at another elementary school will plant seeds for a blue lupine garden near the local airport.

SCIENCE Projects for Home or School

You Can Do It!

Quick and Easy Project

Getting Out the Dirt

Materials
- plastic funnel
- gravel
- sand
- bowl
- water with some soil and leaves in it

Procedure
1. Fill the bottom of a funnel with gravel. Then add a thick layer of sand.
2. Hold the funnel over a bowl. Pour the "dirty" water into it. The water will run out the bottom of the funnel, into the bowl.
3. Observe the water in the bowl. See if you can find the soil and leaves.

Draw Conclusions
How did your funnel filter affect the dirty water? Where might you find this kind of natural filter? How might soil and other substances get into a water supply?

Design Your Own Investigation

Checking for Air Pollution

Is the air in your school or neighborhood polluted? Air pollution often includes bits of ash and dust. If you smear petroleum jelly on the inside of baby-food jars and put them somewhere for several days, bits of pollution may stick to the jelly. Which areas of your school or neighborhood do you think might have this kind of pollution? Write a hypothesis. Then design an experiment, and carry it out to see whether your hypothesis is supported.

Chapter 4 Review and Test Preparation

Vocabulary Review

Use the terms below to complete the sentences. The page numbers tell you where to look in the chapter if you need help.

environment p. 132
ecosystem p. 132
population p. 134
community p. 136
biotic p. 140
abiotic p. 142
diversity p. 146
pollution p. 152

1. A group of maple trees is an example of a _____.

2. The living parts of an ecosystem are _____.

3. All the living and nonliving things in an area interact to form an _____.

4. An ecosystem that includes many kinds of living things has _____.

5. Several kinds of plants and animals living in the same place form a _____.

6. Trash in a stream is one kind of _____.

7. Nonliving factors in an ecosystem are _____ factors.

8. An _____ includes all the living things and nonliving things in an area.

Check Understanding

Write the letter of the best choice.

9. Which of these is an abiotic factor?
   A. lack of food
   B. disease
   C. cold temperatures
   D. introduction of a new plant

10. Which of these is **not** an abiotic factor?
    F. air   H. mushrooms
    G. soil   J. water

11. **MAIN IDEA AND DETAILS** What is the main idea behind planting trees in an area that has been logged?
    A. pollution
    B. habitat restoration
    C. harvesting natural resources
    D. preserving an ecosystem

12. **CAUSE AND EFFECT** Which statement is true about an ecosystem?
    F. Biotic factors are the climate.
    G. The climate affects biotic factors.
    H. Biotic factors cause abiotic factors.
    J. Biotic factors never change.

13. What does the picture show?

  A. abiotic factors
  B. diversity
  C. habitat restoration
  D. pollution

14. Which of these has the greatest effect on an ecosystem?
  F. communities
  G. biotic factors
  H. climate
  J. population

15. Which of these is **never** a result of human actions?
  A. population increases
  B. changes in abiotic factors
  C. natural increase in diversity
  D. pollution

16. Which climate zone is probably shown in the photo?

  F. savanna
  G. taiga
  H. temperate forest
  J. tundra

Inquiry Skills

17. Compare an environment and an ecosystem.

18. A scientist has tracked the migration route of a Yellowstone elk herd every winter for 10 years. The table shows how far south the elk herd has traveled each year. **Draw conclusions** about the elk herd's migration pattern over the 10-year period.

Year	Kilometers Migrated
1990	122
1991	122
1992	130
1993	126
1994	130
1995	133
1996	132
1997	133
1998	133
1999	136

Critical Thinking

19. Imagine that a new kind of animal has suddenly appeared. How might it affect the local ecosystem?

20. A builder has bought land that includes a forest. The builder is planning to put in a housing development.
  Part A Name two possible negative effects the builder and other humans might have on this forest.
  Part B Now name two possible positive effects the builder and other humans might have on this ecosystem.

Chapter 5
Energy Transfer in Ecosystems

Lesson 1 What Are the Roles of Living Things?

Lesson 2 How Do Living Things Get Energy?

Vocabulary

producer
consumer
herbivore
carnivore
omnivore
decomposer
habitat

niche
food chain
prey
predator
food web
energy pyramid

What do YOU wonder?

This lynx must catch and eat hares and many other small animals in order to live. This hare may provide energy for the lynx. Where do hares get the energy they need to live?

Lesson 1

What Are the Roles of Living Things?

Fast Fact

Nothing Fishy About Eating This archer fish is leaping for its prey. It eats insects to get energy for living. Archer fish also hunt by spitting at insects to knock them into the water. Some archer fish are eaten by other animals or die and then decay in the water. In the Investigate, you will find out how decomposers (dee•kuhm•POHZ•erz) help once-living matter decay.

Investigate

Decomposing Bananas

Materials
- 2 slices of banana
- 2 zip-top plastic bags
- spoon
- package of dry yeast
- marker

Procedure

1. Put a banana slice in each bag.

2. Sprinkle $\frac{2}{3}$ spoonful of dry yeast on one banana slice. Yeast is a decomposer, so use the marker to label this bag *D*.

3. Close both bags. Put the bags in the same place.

4. Check both bags every day for a week. Observe and record the changes you see in each bag.

Step 2

Step 4

Draw Conclusions

1. Which banana slice shows more changes? What is the cause of these changes?

2. **Inquiry Skill** Scientists use time relationships to measure progress. How long did it take for your banana slices to begin showing signs of decomposition? How long do you think it would take for your banana slices to completely decompose?

Investigate Further

What will happen if you put flour, instead of yeast, on one banana slice? Write down your prediction, and then try it.

Reading in Science

VOCABULARY
producer p. 166
consumer p. 166
herbivore p. 168
carnivore p. 168
omnivore p. 168
decomposer p. 170

SCIENCE CONCEPTS
▶ how living things use the energy from sunlight
▶ how living things get energy from other living things

READING FOCUS SKILL
MAIN IDEA AND DETAILS
Look for details about the movement of energy among living things.

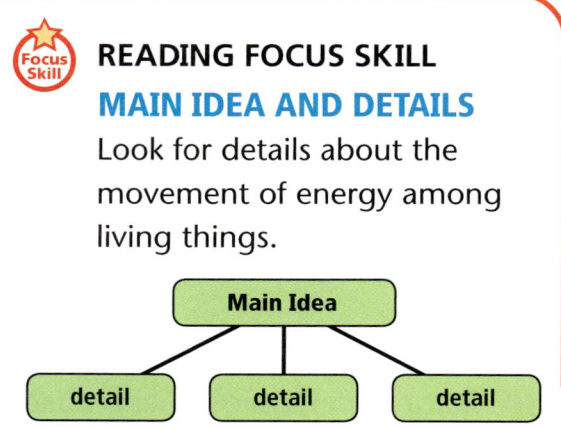

Producers and Consumers

Most living things on Earth get the energy to live from sunlight. Green plants and algae (AL•jee) use energy in sunlight, plus water and carbon dioxide, to make their own food. Any living thing that can make its own food is called a **producer**. Producers can be as small as a tiny moss or as large as a huge redwood tree.

Some animals, such as deer and cattle, get the energy they need to live by eating plants. When these animals eat, the energy stored in the plants moves into the animals' bodies.

Not all animals eat plants. Lions and hawks, for example, get the energy they need by eating other animals.

An animal that eats plants or other animals is called a **consumer**. Consumers can't make their own food, so they must eat other living things.

These plants are using energy in sunlight to produce food. Without sunlight, the plants would die.

Horse

Which animal gets its energy directly from producers? Which one gets its energy from other consumers? Which one gets its energy from both?

Florida panther

Some consumers eat the same kind of food all year. Horses, for example, eat grass during warm weather. During winter, they eat hay, a kind of dried grass.

Other consumers eat different things in different seasons. For example, black bears eat grass in spring. Later on, they might eat birds' eggs. Bears might also dig up tasty roots or eat fish from streams. In fall, bears eat ripe berries.

Florida panthers eat other consumers, but their diet varies. Mostly, panthers consume wild hogs, which are easy for them to catch. Another favorite meal is deer. Panthers also eat rabbits, raccoons, rats, birds, and sometimes, even alligators.

 MAIN IDEA AND DETAILS What is a producer? What is a consumer? Give two examples of each.

Black bear

Kinds of Consumers

Consumers are not all the same. In fact, there are three kinds—herbivores, carnivores, and omnivores.

A **herbivore** is an animal that eats only plants, or producers. Horses are herbivores. So are giraffes, squirrels, and rabbits.

A **carnivore** is an animal that eats only other animals. The Florida panther and the lion are carnivores. A carnivore can be as large as a whale or as small as a frog.

An **omnivore** is an animal that eats both plants and other animals. That is, omnivores eat both producers and other consumers. Bears and hyenas are omnivores. Do any omnivores live in your home?

Producers and all three kinds of consumers can be found living in water. Algae are producers that live in water. They use sunlight to make their own food. Tadpoles, small fish, and other small herbivores eat algae. Larger fish that are carnivores eat the tadpoles. Some animals, including green sea turtles, are omnivores. Green sea turtles eat seaweed, algae, and fish. In fact, algae makes the flesh of the green sea turtle green!

 MAIN IDEA AND DETAILS Name the three kinds of consumers. Give two examples of each.

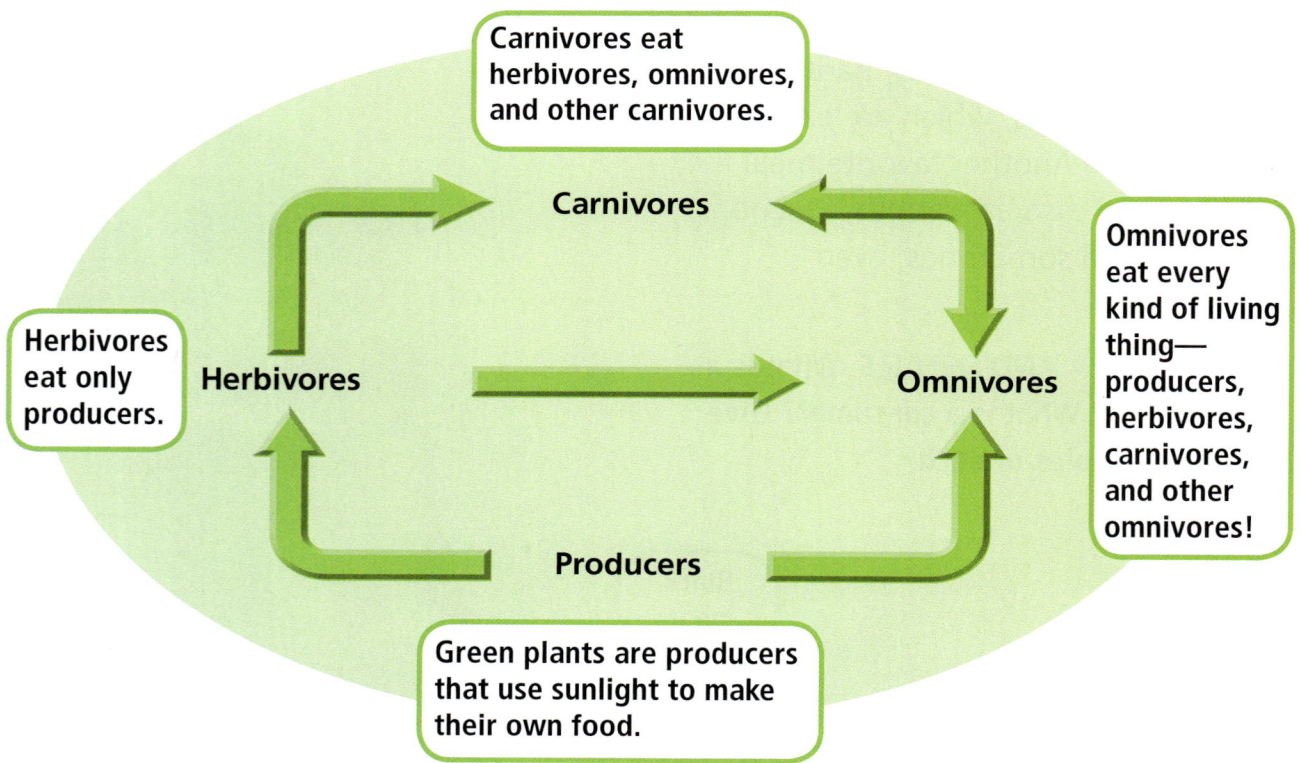

This diagram shows how kinds of consumers get energy to live. The arrows show the direction of energy flow.

168

◄ The jaguar, a carnivore, eats tapirs, river hogs, and other consumers.

Jaguar

Tapir

River hog

▲ The tapir, a herbivore, eats only producers. It eats tender buds and twigs.

▲ River hogs are omnivores. They eat both producers and herbivores.

Jungle bush

▲ This plant is a producer. It makes its own food and provides stored energy for consumers.

Insta-Lab

Who's an Omnivore?

Read the nutrition labels on several food containers. Think about the source of each kind of food. What does the food's source tell about consumers who eat it?

Decomposers

A **decomposer** is a living thing that feeds on wastes and on the remains of dead plants and animals. Decomposers break down wastes into nutrients, substances that are taken in by living things to help them grow. These nutrients become part of the soil. Next, plants take up the nutrients through their roots. Animals eat the plants. When plants and animals die, decomposers break down their bodies into nutrients. This cycle is repeated again and again.

Decomposers come in many shapes and sizes. Some are tiny bacteria that you can see only with a microscope. Other decomposers are as big as mushrooms and earthworms.

Without decomposers, Earth would be covered with dead plants and animals. Instead, decomposers turn wastes into nutrients. They allow living things to recycle nutrients.

 MAIN IDEA AND DETAILS Name two kinds of decomposers, and describe their role in nature.

Sow bugs

Sow bugs are related to lobsters. They help plant matter decay faster than it would without them.

Millipede

In the forest, millipedes chew up dead plant material. Like sow bugs, millipedes aren't insects.

Bracket fungus

The bracket fungus is one of a group of fungi (FUN•jy) that includes mushrooms. Bracket fungi often grow on dead tree trunks and help them decay quickly.

Reading Review

1. **MAIN IDEA AND DETAILS** Copy and complete this graphic organizer.

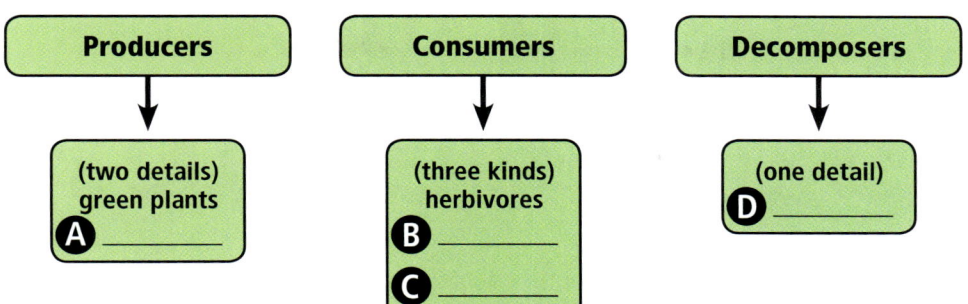

2. **SUMMARIZE** Write two sentences that tell what this lesson is mainly about.

3. **DRAW CONCLUSIONS** How are decomposers consumers?

4. **VOCABULARY** Construct a crossword puzzle, using this lesson's vocabulary words.

Test Prep

5. **Critical Thinking** How do eagles depend on sunlight for their energy?

6. Which term describes a hyena?
   A. carnivore   C. omnivore
   B. herbivore   D. producer

Links

Writing
Narrative Writing
Write a **science fiction story**. Tell about a time when all the producers on Earth disappear. Describe what happens to the consumers.

Math
Solve a Problem
A shrew eats about $\frac{2}{3}$ of its body weight daily. Suppose a child who weighed 30 kilograms (66 lb) could eat $\frac{2}{3}$ of his or her body weight. How many kilograms of food is that?

Health
Eating Decomposers
Find out what vitamins and minerals are in mushrooms. Find healthful recipes that have mushrooms as one of the ingredients.

For more links and activities, go to www.hspscience.com

Lesson 2

How Do Living Things Get Energy?

Fast Fact

Ouch! Only female mosquitoes bite people and other animals. They need the blood to produce eggs. In the Investigate, you will make food chains. You might include a mosquito in yours!

Investigate

Make a Food Chain

Materials
- 8 to 10 blank index cards
- colored pencils or markers
- reference books about animals

Procedure

1. Choose a place where animals live. Some examples are pine forest, rain forest, desert, wetland, and ocean.

2. On an index card, draw a living thing that lives in the place you have chosen. Draw more living things, one kind on each card. Include large animals, small animals, and producers. Look up information about plants and animals if you need help.

Step 2

3. Put your cards in an order that shows what eats what. You might have more than one set of cards. If one of your animals doesn't fit anywhere, trade cards with someone. You can also draw another animal to link two of your cards. For example, you could draw a rabbit to link a grass card and a hawk card.

Step 3

Draw Conclusions

1. Could the same animal fit into more than one set of cards? Explain your answer.

2. **Inquiry Skill** Scientists communicate their ideas in many ways. What do your cards communicate about the relationships of these living things to one another?

Investigate Further

Draw a series of cards in order, with yourself as the last consumer. Compare your role with the roles of other consumers.

Reading in Science

VOCABULARY
habitat p. 174
niche p. 175
food chain p. 176
prey p. 176
predator p. 176
food web p. 178
energy pyramid p. 180

SCIENCE CONCEPTS
▶ how consumers depend on other living things
▶ how energy moves through food chains and food webs

READING FOCUS SKILL
SEQUENCE Look for the order in which things happen.

Habitats

You probably wouldn't see a heron in a desert or a penguin in a swamp. Animals must live in places that meet their needs. A **habitat** is an environment that meets the needs of a living thing. An insect's habitat can be as small as the space under a rock. A migrating bird's habitat can cross a continent.

Many habitats can overlap. For example, the three living things pictured on this page all live in a desert habitat. This desert habitat meets all their needs. Sagebrush grows well here. Sidewinders and tarantulas find many small consumers to eat.

These living things thrive in the desert habitat, even though it's hot and has little water.

The venomous sidewinder eats mice, rats, lizards, and birds. ▶

Sidewinder

◀ Tarantulas are venomous, too. They eat insects, other spiders, and small lizards.

Tarantula

Sagebrush can grow where other plants can't. Sheep and cattle often eat sagebrush in the winter. ▶

Sagebrush

Each living thing in a habitat has a role, or **niche** (NICH). The term *niche* describes how a living thing interacts with its habitat. Part of a living thing's niche is how it gets food and shelter. Its niche also includes how it reproduces, cares for its young, and avoids danger. Each animal has body parts that help it carry out its role. For example, a cat's pointed claws and sharp eyes help it catch its food.

Part of the sidewinder's niche is to eat small animals in its habitat. If all these snakes died, the desert would have too many mice, birds, and lizards. These small animals would eat all the available food and would soon starve. The sidewinder's niche helps keep the number of small desert animals in balance.

 SEQUENCE What would happen next if all the sagebrush disappeared from a desert?

Crab

Lion-fish

Anemone

This coral reef habitat has a balance of producers and consumers. Everything in this picture is a consumer.

Food Chains

Living things depend on one another to live. A **food chain** is the movement of food energy in a sequence of living things. Every food chain starts with producers. Some consumers, such as deer, eat these producers. Then the deer are eaten by other consumers, such as mountain lions. Consumers that are eaten are called **prey**. A consumer that eats prey is a **predator**. Prey are what is hunted. Predators are the hunters.

Some animals in a habitat are prey, while other animals are predators. Predators limit the number of prey animals in a habitat. Wolves are predators of antelope. They keep the population of antelope from increasing too much, so the antelope don't eat all of the producers. Predators often compete for the same prey. This limits the number of predators in a habitat.

 SEQUENCE What would happen next if the number of predators in a habitat increased too much?

A mangrove swamp is one kind of habitat. Special prop roots hold mangrove trees in the muddy soil. Fresh water and salt water mix in this habitat.

Many organisms live in and around the mangrove roots.

Mullets are fish that can live in fresh water or salt water.

Without hawks, the chipmunk population would get very large. The chipmunks would eat all the acorns and then starve.

Acorns provide energy for the chipmunk, which in turn provides energy for the hawk.

An alligator is just one of the predators in a mangrove swamp. Alligators dig burrows for themselves that also provide shelter for other animals during dry times.

Insta-Lab

Chain of Life

Cut white paper into strips that are 2.5 cm (1 in.) by 12.5 cm (5 in.) On each strip, write the name of a producer or a consumer. Then use glue or tape to combine the strips into paper food chains. Which food chains end with you?

Food Webs

A food chain shows how an animal gets energy from one food source. But food chains can overlap. One kind of producer may be food for different kinds of consumers. Some consumers may eat different kinds of food. For example, hawks eat sparrows, mice, and snakes.

Several food chains that overlap form a **food web**. There are food webs in water habitats, too. For example, herons eat snails, fish, and other birds.

On the next page, you can see an ocean food web. It shows that energy moves from plankton, small producers in the ocean, to small shrimp. These shrimp are called *first-level consumers*.

These shrimp then become prey for fish and other *second-level consumers*. They, in turn, are eaten by the biggest fish and mammals in the ocean, called *top-level consumers*.

SEQUENCE What happens after a first-level consumer eats a producer?

Follow several paths in this food web. Begin at the bottom, with a producer, and trace the movement of energy through the web.

Science Up Close

Antarctic Ocean Food Web

This food web begins with energy from the sun. The producers are tiny plants called phytoplankton (FYT•oh•plangk•tuhn). They float near the water's surface because sunlight can't reach deep underwater. No plants grow at the bottom of the ocean. Where would decomposers fit in this food web?

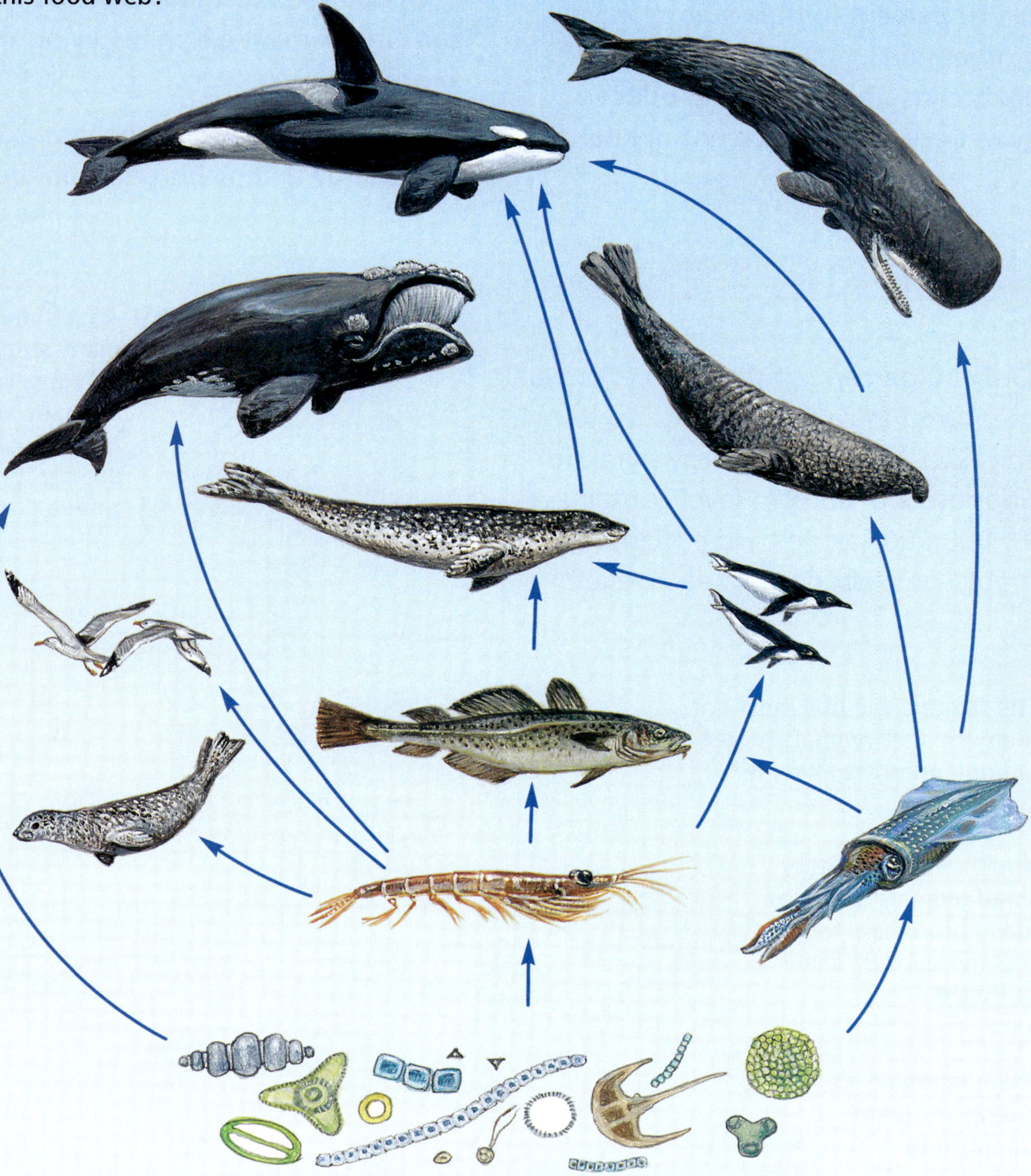

For more links and activities, go to
www.hspscience.com

Energy Pyramids

An **energy pyramid** shows how much energy is passed from one living thing to another along a food chain. Producers form the base of the pyramid. They use about 90 percent of the energy they get from the sun to grow. They store the other 10 percent in their stems, leaves, and other parts.

Next, consumers eat the producers. They get only the 10 percent of energy that the plants stored. These consumers use about 90 percent of the energy they get from the producers to grow and then store the other 10 percent in their bodies. That 10 percent is passed on to the consumers that eat them.

You can see how little energy is passed from one level to the next. That's why consumers must eat many living things in order to live.

 SEQUENCE What happens next to the energy that plants get from the sun?

Math in Science
Interpret Data

Only 10 percent of the food energy, measured in calories, passes up to the next level in an energy pyramid. Suppose the bottom level contains 10,000,000 calories. How many would be passed up to each level?

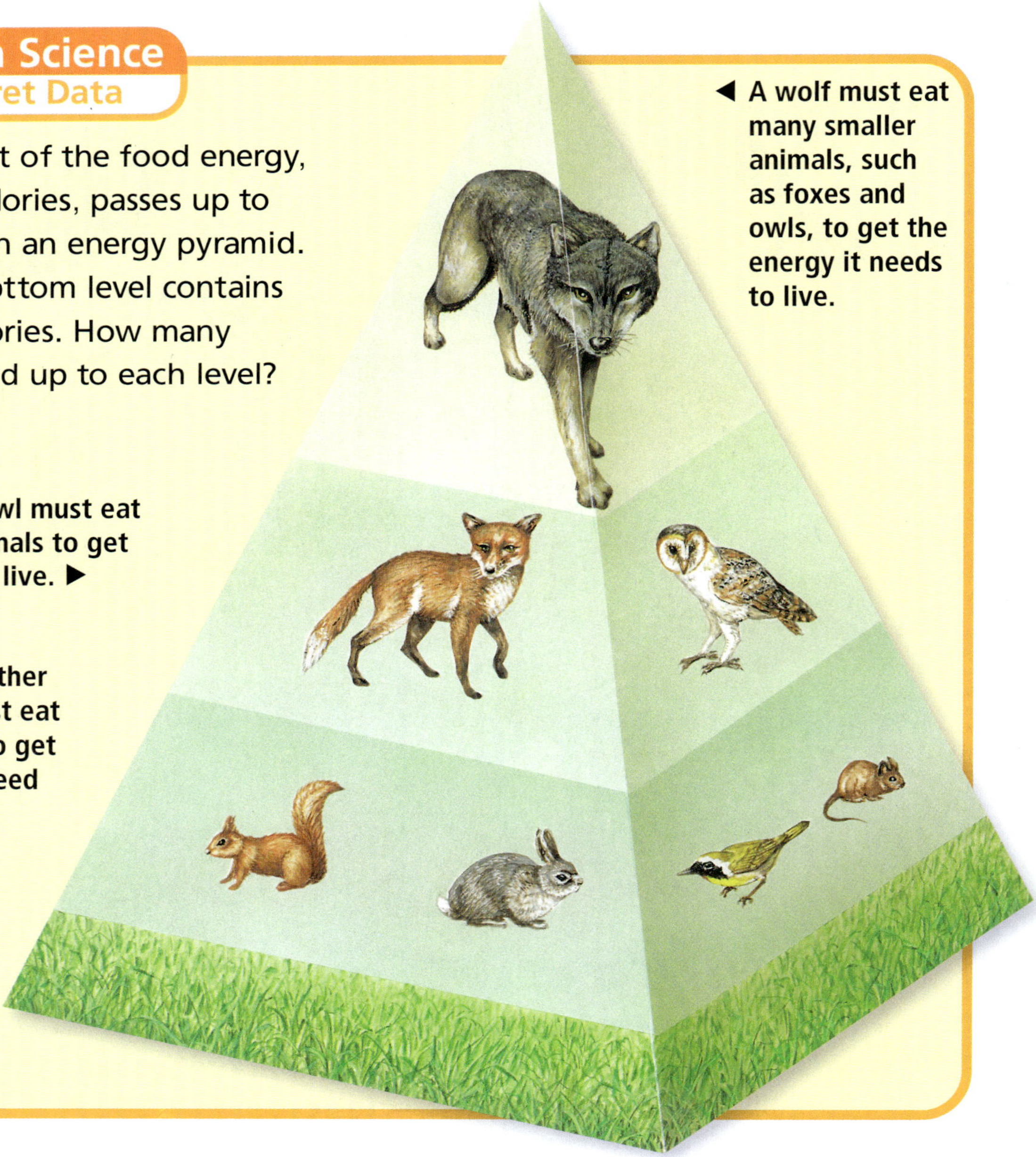

The fox and the owl must eat many smaller animals to get enough energy to live. ▶

Birds, mice, and other small animals must eat many producers to get the energy they need to live. ▶

The bottom of an energy pyramid can include thousands of producers. ▶

◀ A wolf must eat many smaller animals, such as foxes and owls, to get the energy it needs to live.

Reading Review

1. **SEQUENCE** Copy and complete this graphic organizer. Put the living things in order to create a food chain.

woodpecker   hawk   leaves   insect

A ___ → B ___ → C ___ → D ___

2. **SUMMARIZE** Write a summary of this lesson by using the lesson vocabulary terms in a paragraph.

3. **DRAW CONCLUSIONS** How are predators good for prey?

4. **VOCABULARY** Use the vocabulary terms to make a quiz. Then trade quizzes with a partner.

Test Prep

5. **Critical Thinking** How would the deaths of all of one kind of consumer affect a food web?

6. Which of these best shows why deer must eat grass all day long?
   - **A.** diagram
   - **B.** energy pyramid
   - **C.** food chain
   - **D.** food web

Links

Writing
Expository Writing
Write a **description** of ways humans might affect a food web and what would then change. For example, people might clear trees for a housing development or feed the deer in a park.

Math
Solve a Problem
Producers in a field have stored 20,000 calories. Herbivores get 2000 calories by eating the producers. How much energy is available to the next level of the energy pyramid?

Art
Food Chains
Choose any art medium, such as watercolor, charcoal, collage, or torn paper, and show the living things in a food web. (You don't have to show them eating one another!)

For more links and activities, go to www.hspscience.com

On the Prowl

Cameras are helping scientists count jaguars.

A sleek, spotted jaguar sneaks along the thick forest floor. As it passes a fig tree, there is a whirring noise. A flashing light and click follow. A camera has just snapped the cat's photograph.

No person was behind the camera's lens. The camera was triggered by motion and heat from the passing cat.

A Narrowing Range

Scientists from the Wildlife Conservation Society in New York have placed about 30 such cameras in trees throughout the tropical forest of Belize (beh•LEEZ). That is a country in Central America.

The forest is also the site of the world's first jaguar reserve. A reserve is an area set apart for a special purpose. At the reserve in Belize, jaguars are protected and can safely roam.

Belize has a healthy number of jaguars. The wildlife group estimates that about 14 jaguars live within a 143-square-km (55-square-mile) area there. The cameras are helping researchers count the jaguars within certain areas of Belize and in other places where jaguars roam.

"Camera trapping" will help scientists because jaguars are hard to study. Despite the cats' hefty size, their mysterious nature and the thick jungle where they live make them difficult to spot.

A camera snaps a photograph of a passing jaguar.

The map shows how the range of jaguars has changed.

"The cameras help researchers determine how many cats are out there and where they make their homes," jaguar expert Kathleen Conforti told WR.

The researchers will use that information to help protect the endangered animals. They want to conserve, or save, the jaguars' habitat. A habitat is the area where the animal naturally lives.

The actions of people have caused a decline in the animal's range. The cutting down of trees has destroyed some of the jaguar's habitat.

THINK ABOUT IT

1. How might the loss of trees affect how jaguars live?
2. How might equipment such as cameras help protect endangered animals around the world?

What a Roar

- Jaguars, which are carnivorous, can grow up to 1.8 meters (6 ft) long and weigh up to 136 kilograms (300 pounds).
- Jaguars are the third-largest cats, after tigers and lions.
- The cats usually live alone and are very territorial. That means they protect their habitat from other jaguars.
- In Spanish, this cat's name is *el tigre,* which means "the tiger."

Find out more! Log on to www.hspscience.com

SCIENCE Spin from WEEKLY READER
People

WORKING WITH ELEPHANTS

In India, adult Asian elephants have no natural enemies. However, humans have killed many elephants. Now elephants are close to dying out. Raman Sukumar wants to save them.

Sukumar studied how building changes elephant habitats. New dams, roads, and railways force elephants closer to towns. He also studied elephant deaths. He found that illegal hunting has killed many elephants.

Sukumar has found ways to help humans and elephants live together. Areas of the wild are being linked. Elephants can move safely from one area to the next. They don't have to go through farms or towns. Farmers now use different types of fences so that elephants will not eat crops.

Career | Paleontologist

If you like digging, then you may want to become a paleontologist. These scientists study the fossils of ancient animals and plants. As a result, paleontologists can figure out why some species disappeared long ago while other species still exist today.

SCIENCE Projects for Home or School
You Can Do It!

Quick and Easy Project

Energy Pyramid

Materials
- scrap paper
- ruler
- large sheet of white paper
- colored pencils

Procedure

1. Identify producers, herbivores, carnivores, and omnivores that live in your area. List them on scrap paper.
2. Use the ruler to draw a large pyramid on the white paper. Divide the pyramid into three or four levels, depending on the kinds of living things you have identified.
3. Arrange some or all of these living things on your energy pyramid. Draw only one animal at the top level, ten at the next level, and so on.

Draw Conclusions

Do all the things from your list fit into your pyramid? If not, why not? If you lived in a different kind of habitat—for example, a desert or a seashore—how would your energy pyramid change?

Design Your Own Investigation

Carnivore or Herbivore?

Identify several insects of the same kind, such as caterpillars or ants, from your area. Design an experiment to determine if this kind of insect is a herbivore, a carnivore, or an omnivore. For example, you might give the insects a choice of foods and see which foods they eat. Be sure to use safety precautions. Release the insects when the experiment is over.

Chapter 5 Review and Test Preparation

Vocabulary Review

Use the terms below to complete the sentences. The page numbers tell you where to look in the chapter if you need help.

producers p. 166
consumer p. 166
omnivores p. 168
decomposers p. 170
niche p. 175
predators p. 176
food chain p. 176
energy pyramid p. 180

1. An animal that eats other living things is a _____.

2. Nutrients would be lost without _____.

3. The animals at the top of a food chain are always _____.

4. The kind of food that an animal eats is part of its _____.

5. Animals that eat both producers and other consumers are _____.

6. Herbivores and omnivores both eat _____.

7. A food web shows relationships among living things more accurately than a _____.

8. The loss of energy along a food chain is shown in an _____.

Check Understanding

Write the letter of the best choice.

9. Which of these must a pond food chain have?
   A. algae   C. tiny fish
   B. sunlight   D. whales

10. **MAIN IDEA AND DETAILS** Which term includes herbivores, carnivores, and omnivores?
    F. consumers   H. prey
    G. predators   J. producers

11. How much energy is used at each level of the energy pyramid and not passed on?
    A. 10 percent   C. 80 percent
    B. 20 percent   D. 90 percent

12. Which of the following do herbivores eat?
    F. consumers   H. predators
    G. omnivores   J. producers

13. What is shown below?

   A. niche   C. habitat
   B. food chain   D. food web

186

14. What are robins, which eat worms and insects?
- **F.** carnivores
- **G.** herbivores
- **H.** omnivores
- **J.** prey

15. Antelopes are herbivores. What other term describes them?
- **A.** omnivores
- **B.** predators
- **C.** prey
- **D.** producers

16. SEQUENCE What is the first organism on a food chain?
- **F.** a consumer
- **G.** a decomposer
- **H.** a producer
- **J.** a predator

Inquiry Skills

17. Compare a carnivore and a predator. How are these living things the same? How are they different?

18. While hiking with your family, you follow a trail that leads past many dead plants. Even the trees seem to be dying. The soil is very dry. What can you **infer** is happening to the consumers in this area?

Critical Thinking

19. Which of these could survive without being part of a food chain— a strawberry plant, a chicken, or a dog? Explain your answer.

20. Different types of diagrams are used to show the relationships among living things. Study the diagram below.

Part A Would this diagram be correct if there were two snakes at the top? Explain your answer.
Part B How is this diagram different from a food chain?

References

Contents

Health Handbook

Your Skin R1

Your Digestive System R2

Your Circulatory System R4

Your Skeletal System R6

Your Muscular System R8

Your Senses R10

Your Immune System R12

Staying Healthy R14

Reading in Science Handbook

Identify the Main Idea and Details R16

Compare and Contrast R18

Identify Cause and Effect R20

Sequence R22

Summarize R24

Draw Conclusions R26

Math in Science Handbook — R28

Safety in Science R36

Glossary R37

Index R46

Health Handbook

Your Skin

Your skin is your body's largest organ. It provides your body with a tough protective covering. It protects you from disease. It provides your sense of touch, which allows you to feel pressure, textures, temperature, and pain. Your skin also produces sweat to help control your body temperature. When you play hard or exercise, your body produces sweat, which cools you as it evaporates. The sweat from your skin also helps your body get rid of extra salt and other wastes.

▼ The skin is the body's largest organ.

Epidermis
Many layers of dead skin cells form the top of the epidermis. Cells in the lower part of the epidermis are always making new cells.

Oil Gland
Oil glands produce oil that keeps your skin soft and smooth.

Hair Follicle
Each hair follicle has a muscle that can contract and make the hair "stand on end."

Pore
These tiny holes on the surface of your skin lead to your dermis.

Sweat Gland
Sweat glands produce sweat, which contains water, salt, and various wastes.

Dermis
The dermis is much thicker than the epidermis. It is made up of tough, flexible fibers.

Fatty Tissue
This tissue layer beneath the dermis stores food, provides warmth, and attaches your skin to the bone and muscle below.

Caring for Your Skin

- To protect your skin and to keep it healthy, you should wash your body, including your hair and your nails, every day. This helps remove germs, excess oils and sweat, and dead cells from the epidermis, the outer layer of your skin. Because you touch many things during the day, you should wash your hands with soap and water frequently.

- If you get a cut or scratch, you should wash it right away and cover it with a sterile bandage to prevent infection and promote healing.

- Protect your skin from cuts and scrapes by wearing proper safety equipment when you play sports or skate, or when you're riding your bike or scooter.

Your Digestive System

Your digestive system is made up of connected organs. It breaks down the food you eat and disposes of the leftover wastes your body does not need.

Mouth to Stomach

Digestion begins when you chew your food. Chewing your food breaks it up and mixes it with saliva. When you swallow, the softened food travels down your esophagus to your stomach, where it is mixed with digestive juices. These are strong acids that continue the process of breaking your food down into the nutrients your body needs to stay healthy. Your stomach squeezes your food and turns it into a thick liquid.

Small Intestine and Liver

Your food leaves your stomach and goes into your small intestine. This organ is a long tube just below your stomach. Your liver is an organ that sends bile into your small intestine to continue the process of digesting fats in the food. The walls of the small intestine are lined with millions of small, finger-shaped bumps called villi. Tiny blood vessels in these bumps absorb nutrients from the food as it moves through the small intestine.

Large Intestine

When the food has traveled all the way through your small intestine, it passes into your large intestine. This last organ of your digestive system absorbs water from the food. The remaining wastes are held there until you go to the bathroom.

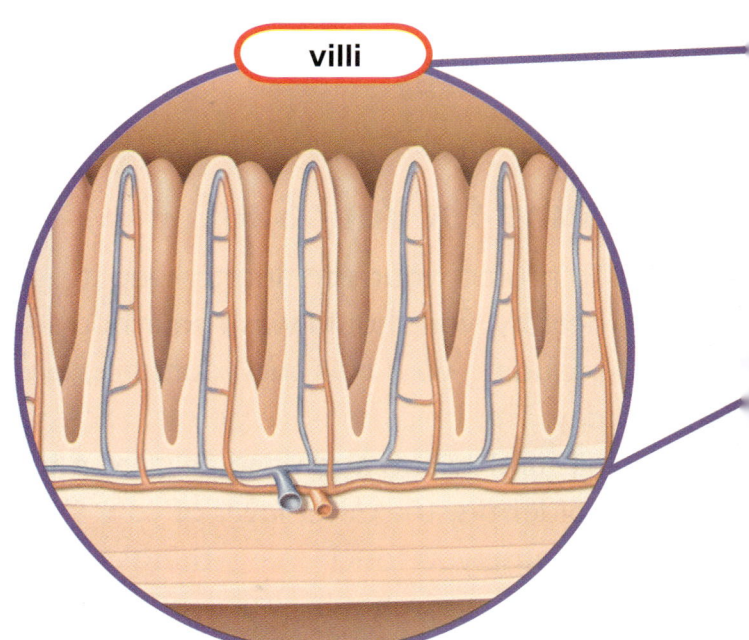

villi

Health Handbook

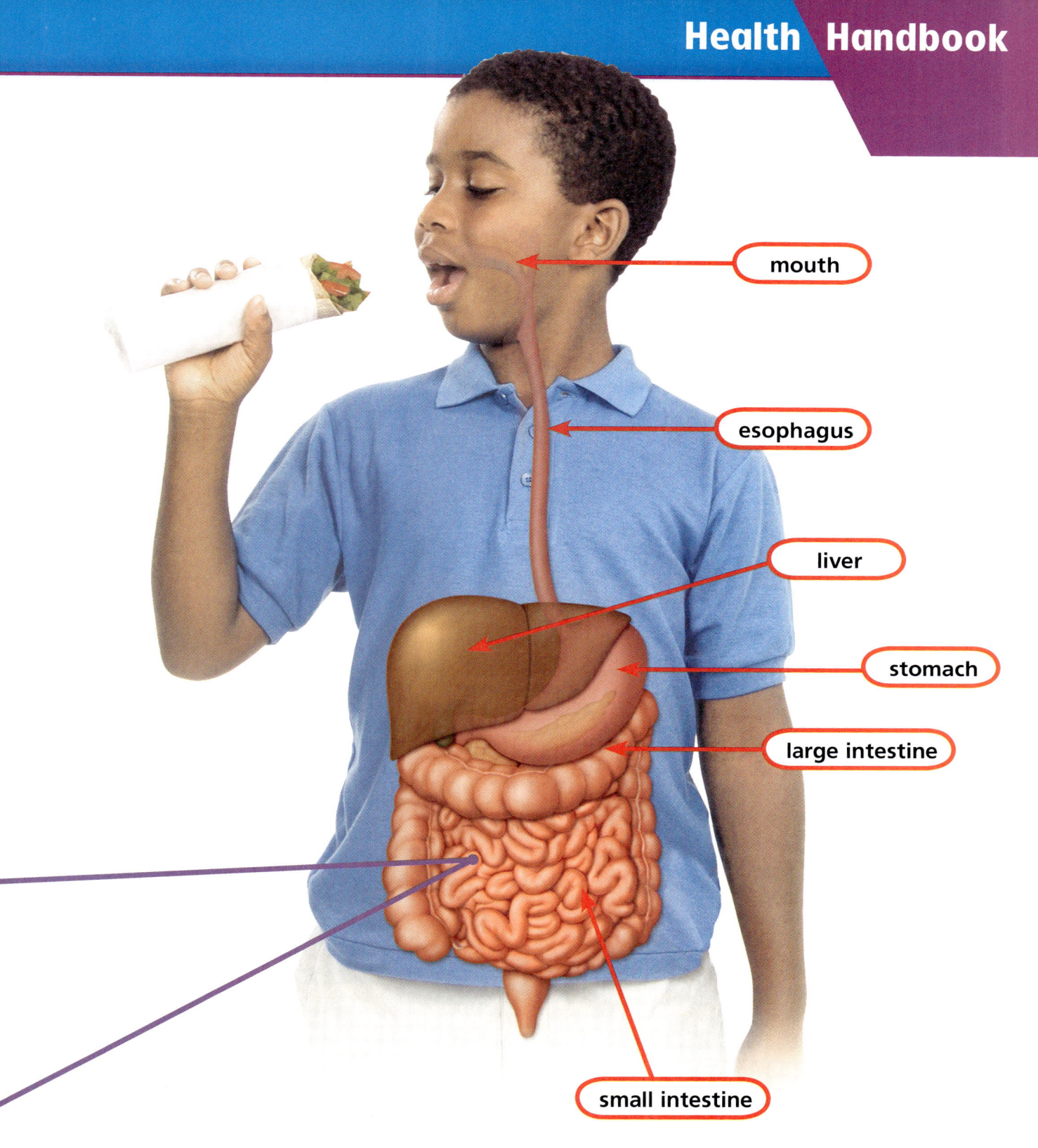

Your Circulatory System

Your circulatory system carries to every cell in your body the nutrients your digestive system takes from food and the oxygen your lungs take from the air you breathe. As your blood moves throughout your body, it also helps your body fight infections, control your temperature, and remove wastes from your cells.

Your Heart and Blood Vessels

Your heart is the organ that pumps your blood through your circulatory system. Your heart is a strong muscle that beats continuously. As you exercise, your heart adjusts itself to beat faster to deliver the energy and oxygen your muscles need to work harder.

Blood from your heart is pumped through veins into your lungs, where it releases carbon dioxide and picks up oxygen. Your blood then travels back to your heart to be pumped through your arteries to every part of your body.

Your Blood

The blood in your circulatory system is a mixture of fluids and specialized cells. The watery liquid part of your blood is called plasma. Plasma allows the cells in your blood to move through your blood vessels to every part of your body. It also plays an important role in helping your body control your temperature.

Health Handbook

Blood Cells

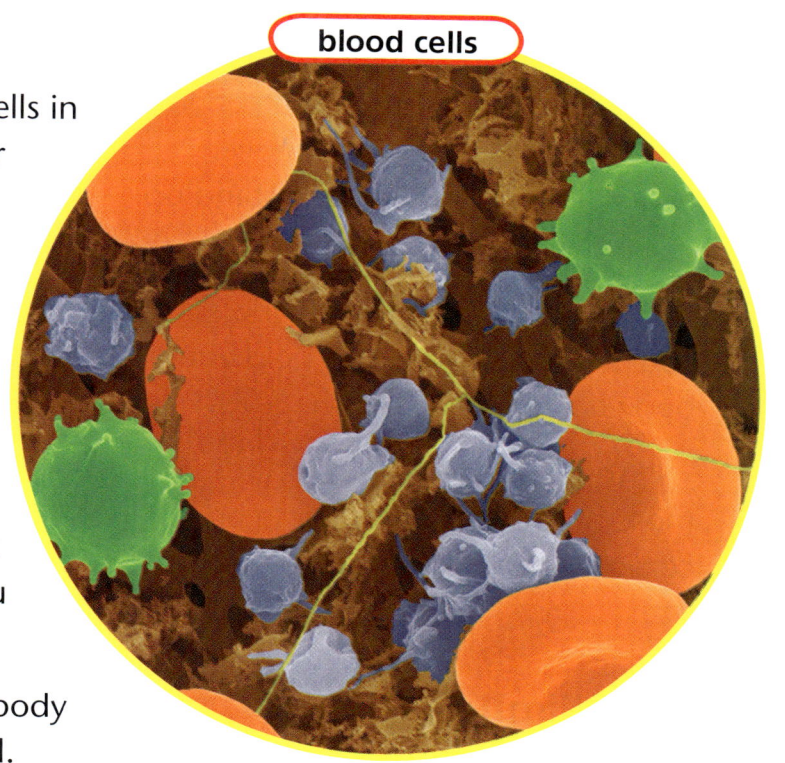

There are three main types of cells in your blood. Each type of cell in your circulatory system plays a special part in keeping your body healthy and fit.

Red Blood Cells are the most numerous cells in your blood. They carry oxygen from your lungs throughout your body. They also carry carbon dioxide back to your lungs from your cells, so you can breathe it out.

White Blood Cells help your body fight infections when you become ill.

Platelets help your body stop bleeding when you get a cut or other wound. Platelets clump together as soon as you start to bleed. The sticky clump of platelets traps red blood cells and forms a blood clot. The blood clot hardens to make a scab that seals the cut and lets your body begin healing the wound.

Caring for Your Circulatory System

- Eat foods that are low in fat and high in fiber. Fiber helps take away substances that can lead to fatty buildup in your blood vessels.
- Eat foods high in iron to help your red blood cells carry oxygen.
- Drink plenty of water to help your body replenish your blood.
- Avoid contact with another person's blood.
- Exercise regularly to keep your heart strong.
- Never smoke or use tobacco.

R5

Your Skeletal System

Your skeletal system includes all of the bones in your body. These strong, hard parts of your body protect your internal organs, help you move, and allow you to sit and to stand up straight.

Your skeletal system works with your muscular system to hold your body up and to give it shape.

Your skeletal system includes more than 200 bones. These bones come in many different shapes and sizes.

Your Skull

The wide flat bones of your skull fit tightly together to protect your brain. The bones in the front of your skull give your face its shape and allow the muscles in your face to express your thoughts and feelings.

Your Spine

Your spine, or backbone, is made up of nearly two dozen small, round bones. These bones fit together and connect your head to your pelvis. Each of these bones, or vertebrae (VUHR•tuh•bree), is shaped like a doughnut with a small round hole in the center. Your spinal cord is a bundle of nerves that carries information to and from your brain and the rest of your body. Your spinal cord runs from your brain down your back to your hips through the holes in your vertebrae. There are soft, flexible disks of cartilage between your vertebrae. This allows you to bend and twist your spine. Your spine, pelvis, and leg bones work together to allow you to stand, sit, or move.

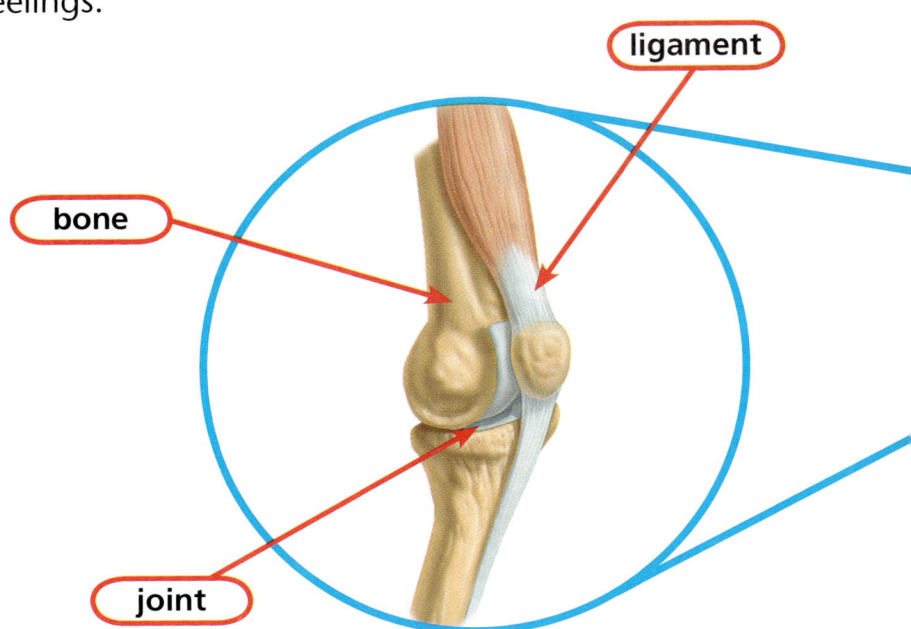

Health Handbook

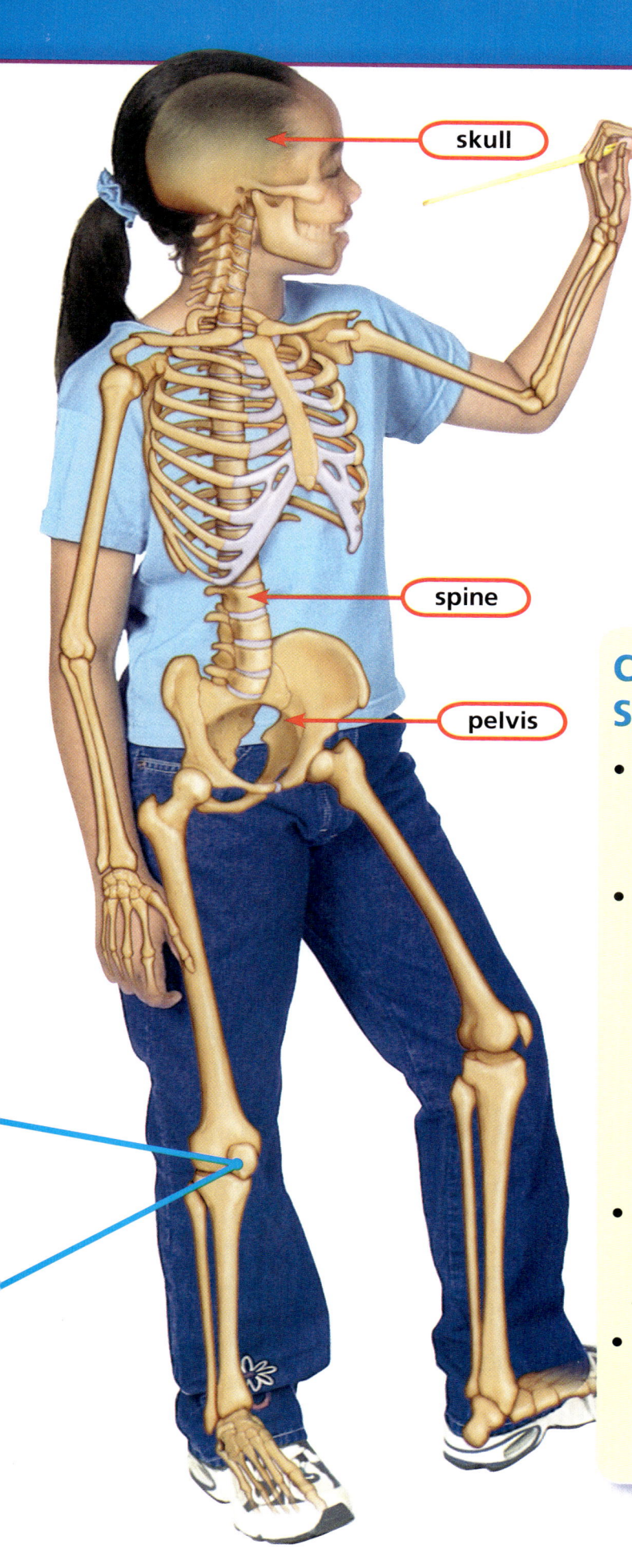

Caring for Your Skeletal System

- Always wear a helmet and proper safety gear when you play sports, skate, or ride a bike or a scooter.

- Your bones are made mostly of calcium and other minerals. To keep your skeletal system strong and to help it grow, you should eat foods that are high in calcium like milk, cheese, and yogurt. Dark green, leafy vegetables like broccoli, spinach, and collard greens are also good sources of calcium.

- Exercise to help your bones stay strong and healthy. Get plenty of rest to help your bones grow.

- Stand and sit with good posture. Sitting slumped over puts strain on your muscles and on your bones.

R7

Your Muscular System

A muscle is a body part that produces movement by contracting and relaxing. All of the muscles in your body make up the muscular system.

Voluntary and Involuntary Muscles

Voluntary Muscles are the muscles you use to move your arms and legs, your face, head, and fingers. You can make these muscles contract or relax to control the way your body moves.

Involuntary Muscles are responsible for movements you usually don't see or control. These muscles make up your heart, your stomach and digestive system, your diaphragm, and the muscles that control your eyelids. Your heart beats and your diaphragm powers your breathing without your thinking about them. You cannot stop the action of these muscles.

How Muscles Help You Move

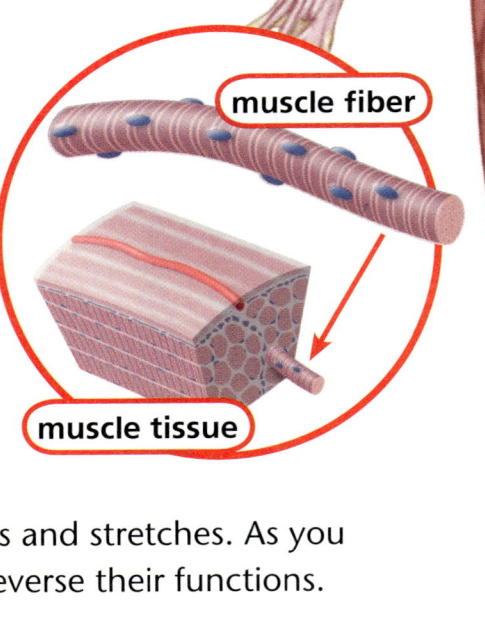

All muscles pull when they contract. Moving your body in more than one direction takes more than one muscle. To reach out with your arm or to pull it back, you use a pair of muscles. As one muscle contracts to extend your arm, the other relaxes and stretches. As you pull your arm back, the muscles reverse their functions.

Your muscles let you do many kinds of things. The large muscles in your legs allow you to walk and run. Tiny muscles in your face allow you to smile.

Health Handbook

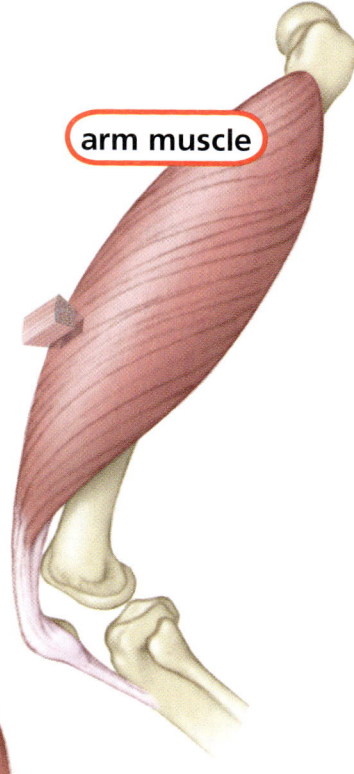

arm muscle

Your Muscles and Your Bones

The muscles that allow you to move your body work with your skeletal system. Muscles in your legs that allow you to kick a ball or ride a bicycle pull on the bones and joints of your legs and lower body. Your muscles are connected to your skeletal system by strong, cordlike tissues called tendons.

Your Achilles tendon just above your heel connects your calf muscles to your heel bone. When you contract those muscles, the tendon pulls on the heel bone and allows you to stand on your toes, jump, or push hard on your bicycle's pedals.

Caring for Your Muscular System

- Always stretch and warm your muscles up before exercising or playing sports. Do this by jogging or walking for at least ten minutes. This brings fresh blood and oxygen into your muscles and helps prevent injury or pain.

- Eat a balanced diet of foods to be sure your muscles have the nutrients they need to grow and remain strong.

- Drink plenty of water when you exercise or play sports. This helps your blood remove wastes from your muscles and helps you build endurance.

- Always cool down after you exercise. Walk or jog slowly for five or ten minutes to let your heartbeat slow and your breathing return to normal. This helps you avoid pain and stiffness after your muscles work hard.

- Stop exercising if you feel pain in your muscles.

- Get plenty of rest before and after you work your muscles hard. They need time to repair themselves and recover from working hard.

Your Senses

Your Eyes and Vision

Your eyes allow you to see light reflected by the things around you. This diagram shows how an eye works. Light enters through the clear outer surface called the cornea. It passes through the pupil. The lens bends the incoming light to focus it on the retina. The retina sends nerve signals along the optic nerve. Your brain uses the signals to form an image. This is what you "see."

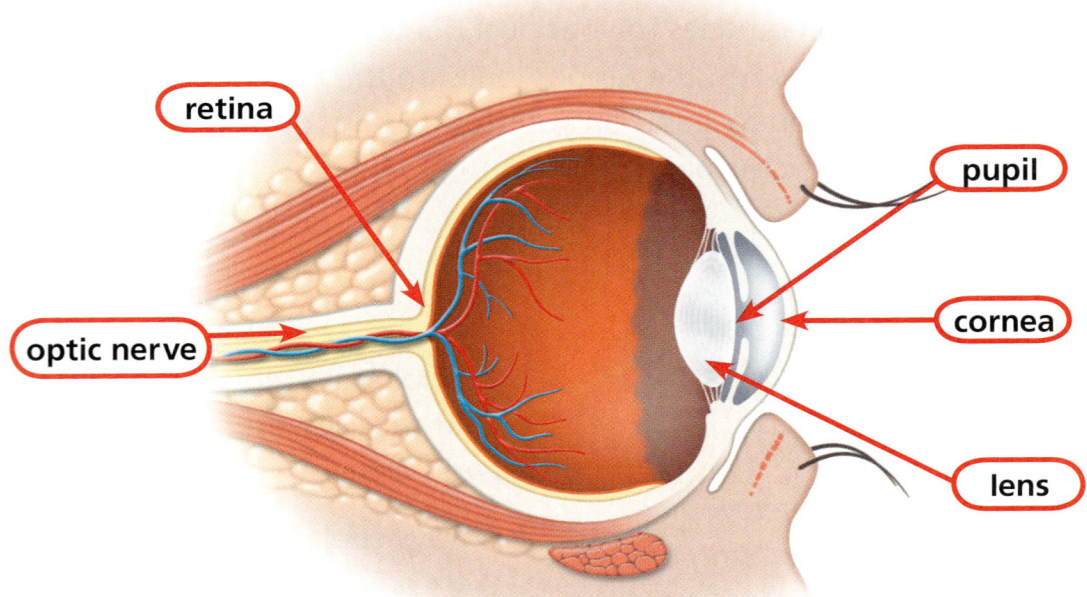

Caring for Your Eyes

- You should have a doctor check your eyesight every year. Tell your parents or your doctor if your vision becomes blurry or if you are having headaches or pain in your eyes.

- Never touch or rub your eyes.

- Protect your eyes by wearing safety goggles when you use tools or play sports.

- Wear swim goggles to protect your eyes from chlorine or other substances in the water.

- Wear sunglasses to protect your eyes from very bright light. Looking directly at bright light or at the sun can damage your eyes permanently.

Health Handbook

Your Ears and Hearing

Sounds travel through the air in waves. When some of those waves enter your ear you hear a sound. This diagram shows the inside of your ear.

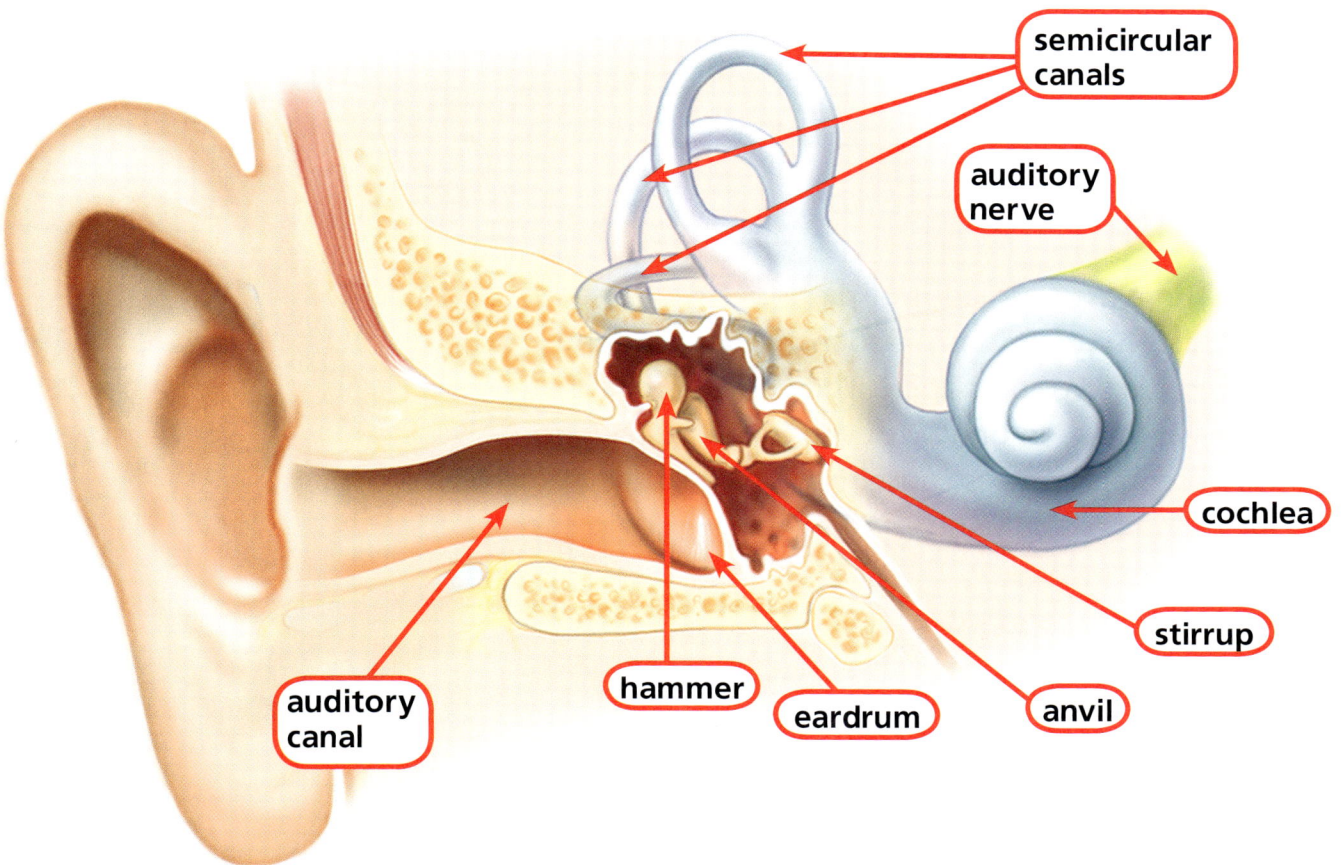

Caring for Your Ears

- Never put anything in your ears.
- Wear a helmet that covers your ears when you play sports.
- Keep your ears warm in winter.
- Avoid loud sounds and listening to loud music.
- Have your ears checked by a doctor if they hurt or leak fluid or if you have any loss of hearing.
- Wear earplugs when you swim. Water in your ears can lead to infection.

Your Immune System

Pathogens and Illness

You may know someone who had a cold or the flu this year. These illnesses are caused by germs called pathogens. Illnesses spread when pathogens move from one person to another.

Types of Pathogens

There are four kinds of pathogens—viruses, bacteria, fungi, and protozoans. Viruses are the smallest kind of pathogen. They are so small that they can be seen only with very powerful electron microscopes. Viruses cause many types of illness, including colds, the flu, and chicken pox. Viruses cannot reproduce by themselves. They must use living cells to reproduce.

Bacteria are tiny single-cell organisms that live in water, in the soil, and on almost all surfaces. Most bacteria can be seen only with a microscope. Not all bacteria cause illness. Your body needs some types of bacteria to work well.

The most common type of fungus infection is athlete's foot. This is a burning, itchy infection of the skin between your toes. Ringworm is another skin infection caused by a fungus. It causes itchy round patches to develop on the skin.

Protozoans are the fourth type of pathogen. They are single-cell organisms that are slightly larger than bacteria. They can cause disease when they grow in food or drinking water.

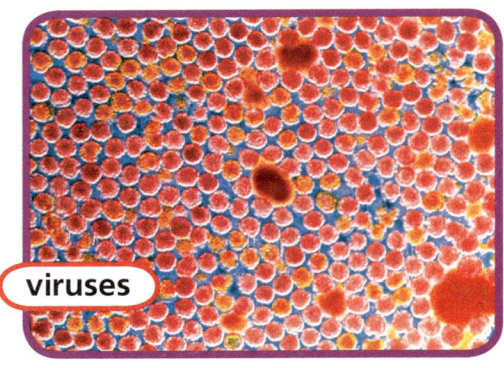
viruses

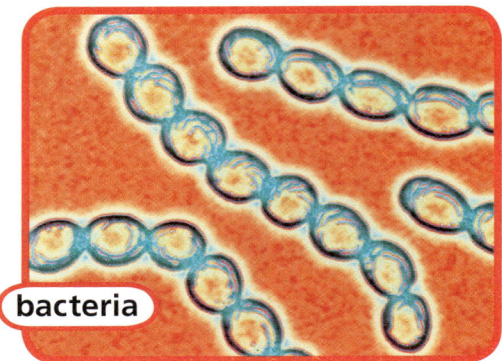
bacteria

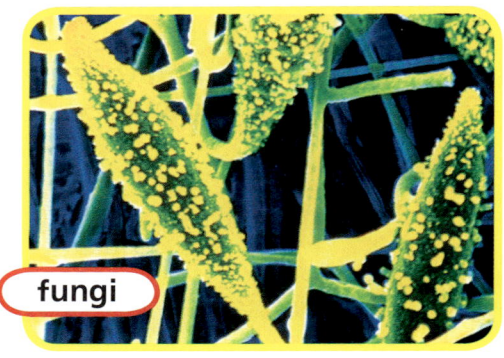
fungi

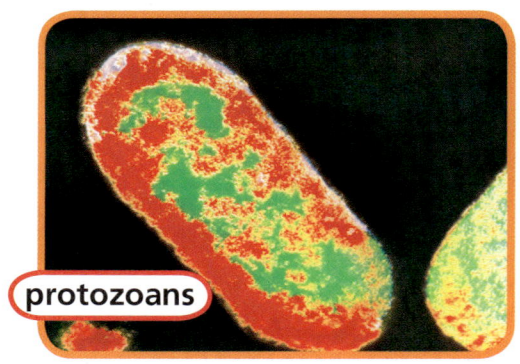
protozoans

Health Handbook

Fighting Illness

Pathogens that can make you ill are everywhere. When you become ill, a doctor may be able to treat you. You also can practice healthful habits to protect yourself and others from the spread of pathogens and the illnesses they can cause.

The best way to avoid spreading pathogens is to wash your hands with warm water and soap. This floats germs off of your skin. You should wash your hands often. Always wash them before and after eating, after handling animals, and after using the bathroom. Avoid touching your mouth, eyes, and nose. Never share hats, combs, cups, or drinking straws. If you get a cut or scrape, pathogens can enter your body. It is important to wash cuts and scrapes carefully with soap and water. Then cover the injury with a sterile bandage.

When you are ill, you should avoid spreading pathogens to others. Cover your nose and mouth when you sneeze or cough. Don't share anything that has touched your mouth or nose. Stay home from school until an adult or your doctor tells you that you are well enough to go back.

Even though pathogens are all around, most people become ill only once in a while because the body has systems that protect it from pathogens. These defenses keep pathogens from entering your body.

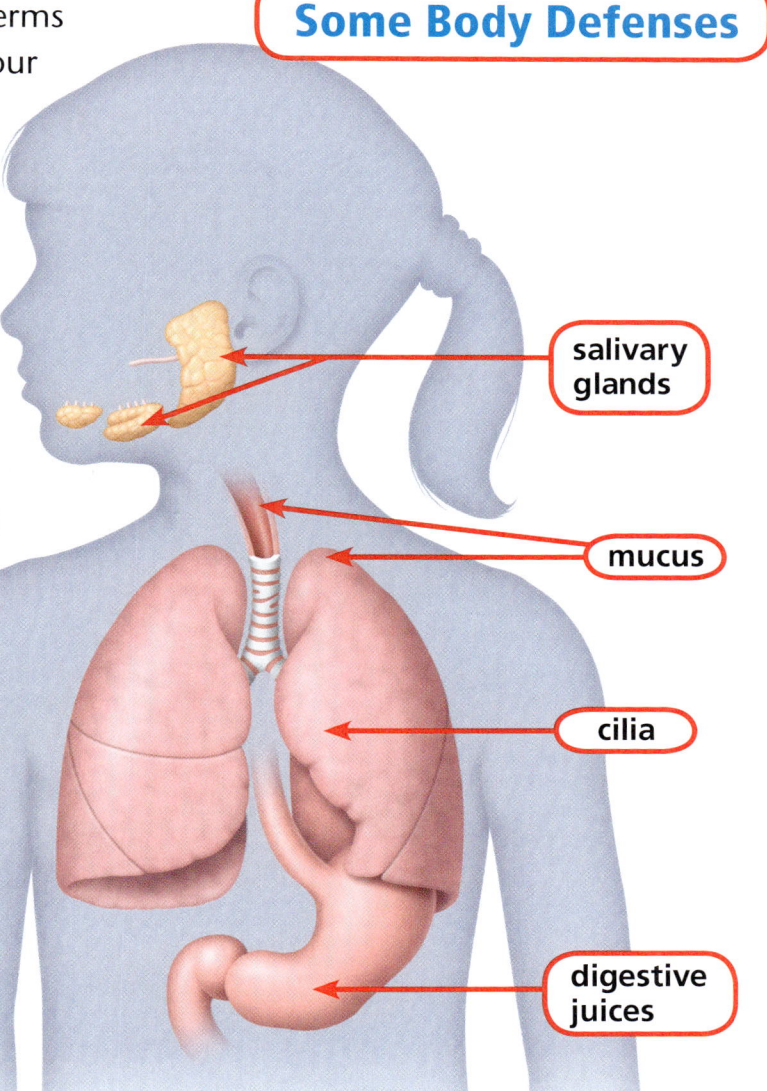

Some Body Defenses
- salivary glands
- mucus
- cilia
- digestive juices

Staying Healthy

Eat a Balanced Diet

Eating the foods that your body needs to grow and fight illness is the most important thing you can do to stay healthy. A balanced diet of healthful foods gives your body energy. Your body's systems need nutrients to function properly and work together.

Choosing unhealthful foods can cause you to gain excess weight and to lack energy. Inactivity and poor food choices can lead to your becoming ill more frequently. Unhealthful foods can also cause you to develop noncommunicable diseases. Unlike communicable diseases, which are caused by germs, these illnesses occur because your body systems are not working right.

Health Handbook

Exercise Regularly

Exercise keeps your body healthy. Regular exercise helps your heart, lungs, and muscles stay strong. It helps your body digest food. It also helps your body fight disease. Exercising to keep your body strong also helps prevent injury when you play sports.

Exercise allows your body to rest more effectively. Getting enough sleep prepares your body for the next day. It allows your muscles and bones to grow and recover from exercise. Resting also helps keep your mind alert so you can learn and play well.

Identify the Main Idea and Details

Many of the lessons in this science book are written so that you can understand main ideas and the details that support them. You can use a graphic organizer like this one to show a main idea and details.

Main Idea: The most important idea of a selection

- **Detail:** Information that tells more about the main idea
- **Detail:** Information that tells more about the main idea
- **Detail:** Information that tells more about the main idea

Tips for Identifying the Main Idea and Details

- To find the main idea, ask *What is this mostly about?*
- Remember that the main idea is not always stated in the first sentence of a passage.
- Look for details that answer questions such as *who, what, where, when, why,* and *how*. Use pictures as clues to help you.

Here is an example.

Main Idea

An environment that meets the needs of a living thing is called its habitat. Some habitats are as big as a whole forest. This is often true for birds that fly from place to place. Some habitats are very small. For example, fungi might grow only in certain places on a forest floor.

Detail

Here is what you could record in the graphic organizer.

Main Idea: An environment that meets the needs of a living thing is called its habitat.

- **Detail:** Some habitats are as big as a whole forest.
- **Detail:** A bird's habitat might be a whole forest.
- **Detail:** Fungi might grow only in certain places on a forest floor.

Reading in Science Handbook

More About Main Idea and Details

Sometimes the main idea is not at the beginning of a passage. If the main idea is not stated, it can be understood from the details. Look at the graphic organizer. What do you think the main idea is?

Main Idea:

- **Detail:** Green plants are the producers in a food chain. They make their own food.
- **Detail:** Consumers make up the next level of a food chain. They eat plants and other living things for energy.
- **Detail:** Decomposers are the next level. They feed on the wastes of consumers or on their remains.

A paragraph's main idea may be supported by details of different types. In this paragraph, identify whether the details give reasons, examples, facts, steps, or descriptions.

> A group of the same species living in the same place at the same time is called a population. A forest may have populations of several different kinds of trees. Trout may be one of several populations of fish in a stream. Deer may be one population among many in a meadow.

Skill Practice

Read the following paragraph. Use the Tips for Identifying the Main Idea and Details to answer the questions.

> Animals do not get their energy directly from the sun. Many eat plants. The plants use sunlight to make food. Animals that don't eat plants still depend on the energy of sunlight. They eat animals that eat plants. The sun is the main source of energy for all living things.

1. What is the main idea of the paragraph?
2. What supporting details give more information about the main idea?
3. What details answer any of the questions *who, what, where, when, why,* and *how*?

R17

Compare and Contrast

Some lessons are written to help you see how things are alike or different. You can use a graphic organizer like this one to compare and contrast.

Topic: Name the two things you are comparing and contrasting.

Alike
List ways the things are alike.

Different
List ways the things are different.

Tips for Comparing and Contrasting

- To compare, ask *How are the people, places, objects, ideas, or events alike?*
- To contrast, ask *How are the people, places, objects, ideas, or events different?*
- When you compare, look for signal words and phrases such as *similar, alike, both, the same as, too,* and *also.*
- When you contrast, look for signal words and phrases such as *unlike, different, however, yet,* and *but.*

Here is an example.

Compare

Mars and Venus are the two planets closest to Earth. They are known as inner planets. Venus and Earth are about the same size, but Mars is a little smaller. Venus does not have any moons. However, Mars has two moons.

Contrast

Here is what you could record in the graphic organizer.

Topic: Mars and Venus

Alike
Both are inner planets.
Are the planets closest to Earth.

Different
Mars is smaller than Venus.
Mars has two moons.

Reading in Science Handbook

More About Compare and Contrast

You can better understand new information about things when you know how they are alike and how they are different. Use the graphic organizer from page R18 to sort the following items of information about Mars and Venus.

Mars	Venus
Mars is the fourth planet from the sun.	Venus is the second planet from the sun.
A year on Mars is 687 Earth days.	A year on Venus is 225 Earth days.
Mars has a diameter of 6794 kilometers.	Venus has a diameter of 12,104 kilometers.
The soil on Mars is a dark reddish brown.	Venus is dry and has a thick atmosphere.

Sometimes a paragraph compares and contrasts more than one topic. In the following paragraph, one topic being compared and contrasted is underlined. Find the second topic being compared and contrasted.

> Radio telescopes and optical telescopes are two types of telescopes that are used to observe objects in space. A radio telescope collects radio waves with a large, bowl-shaped antenna. Optical telescopes use light. There are two types of optical telescopes. A refracting telescope uses lenses to magnify an object and a reflecting telescope uses a curved mirror to magnify an object.

Skill Practice

Read the following paragraph. Use the Tips for Comparing and Contrasting to answer the questions.

> Radio telescopes and optical telescopes work in the same way. However, optical telescopes collect and focus light, while radio telescopes collect and focus invisible radio waves. Radio waves are not affected by clouds and poor weather. Computers can make pictures from data collected by radio telescopes.

1. How are radio and optical telescopes alike? Different?
2. What are two compare and contrast signal words in the paragraph?

Identify Cause and Effect

Some of the lessons in this science book are written to help you understand why things happen. You can use a graphic organizer like this one to show cause and effect.

Cause		Effect
A cause is an action or event that makes something happen.	→	An effect is what happens as a result of an action or event.

Tips for Identifying Cause and Effect

- To find an effect, ask *What happened?*
- To find a cause, ask *Why did this happen?*
- Remember that actions and events can have more than one cause or effect.
- Look for signal words and phrases such as *because* and *as a result* to help you identify causes and effects.

Here is an example.

Cause

Effect

A pulley is a simple machine. It helps us do work. It is made up of a rope or chain and a wheel around which the rope fits. When you pull down on one rope end, the wheel turns and the other rope end moves up.

Here is what you could record in the graphic organizer.

Cause		Effect
One rope end is pulled down on a pulley.	→	The wheel of the pulley turns and the other rope end moves up.

Reading in Science Handbook

More About Cause and Effect

Actions and events can have more than one cause or effect. For example, suppose the paragraph on page R20 included a sentence that said *The pulley can be used to raise or lower something that is light in weight.* You could then identify two effects of operating a pulley.

Cause
One rope end is pulled down on a pulley.

Effect
The wheel of the pulley turns and the other rope end moves up.

Effect
Something light in weight is raised or lowered.

Some paragraphs contain more than one cause and effect. In the following paragraph, one cause and its effect are underlined. Find the second cause and its effect.

A fixed pulley and a movable pulley can be put together to make a compound machine. The movable pulley increases your force. As more movable pulleys are added to a system, the force is increased. The fixed pulley changes the direction of your force.

Skill Practice

Read the following paragraph. Use the Tips for Identifying Cause and Effect to help you answer the questions.

A lever can be used to open a paint can. The outer rim of the can is used as the fulcrum. Your hand supplies the effort force. The force put out by the end under the lid is greater than the effort force. As a result, the can is opened.

1. What causes the paint can to open?
2. What is the effect when an effort force is applied?
3. What signal phrase helped you identify the cause and effect in this paragraph?

Sequence

Some lessons in this science book are written to help you understand the order in which things happen. You can use a graphic organizer like this one to show a sequence.

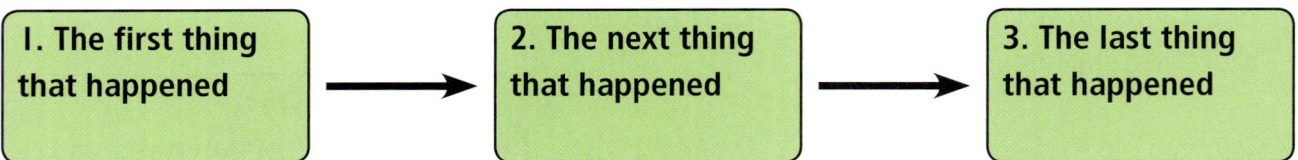

Tips for Understanding a Sequence

- Pay attention to the order in which events happen.
- Recall dates and times to help you understand the sequence.
- Look for signal words such as *first, next, then, last,* and *finally*.
- Sometimes it is helpful to add your own time-order words to help you understand a sequence.

Here is an example.

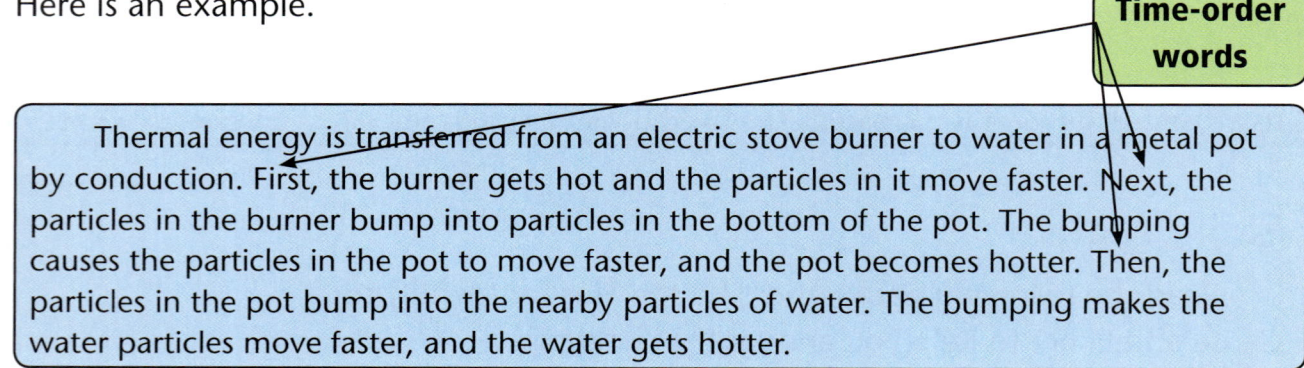

Here is what you could record in the graphic organizer.

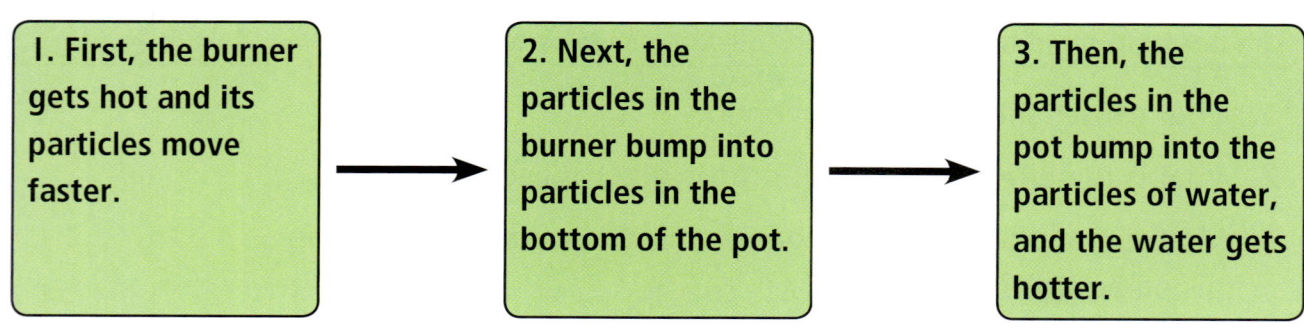

Reading in Science Handbook

More About Sequence

Sometimes information is sequenced by time. For example, an experiment might be done to measure temperature change over time. Use the graphic organizer to sequence the experiment.

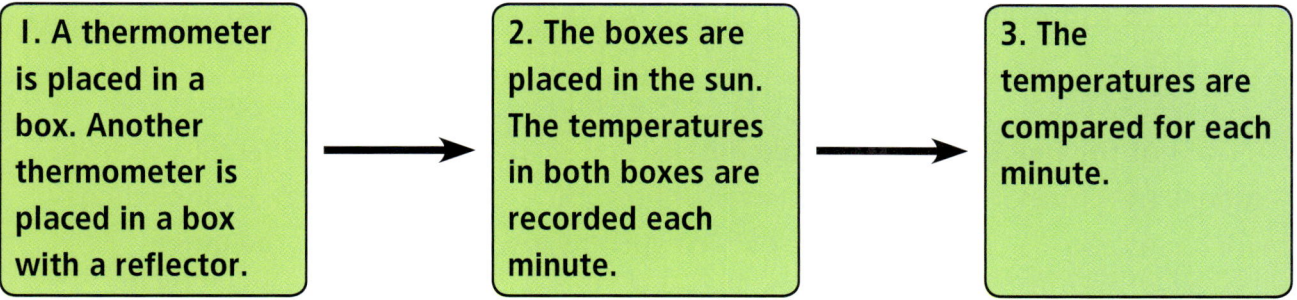

When time-order words are not given, add your own words to help you understand the sequence. In the paragraph below, one time-order word has been included and underlined. How many more time-order words can you add to understand the paragraph's sequence?

> Convection is the transfer of thermal energy in a fluid, a liquid or gas. As the fluid near a hot object gets hot, it expands. The hot fluid is forced up by the cooler, denser fluid around it. As the hot fluid is forced up, it warms the fluid around it. Then, it slowly cools as it sinks.

Skill Practice

Read the following paragraph. Use the Tips for Understanding a Sequence to answer the questions.

> Solar energy can be used to heat water in a home. First, solar panels are placed on the roof of a house. Next, the panels absorb infrared radiation from the sun. Then, the radiation heats the water as it flows through the panels.

1. What is the first thing that happens in the sequence?
2. How many steps are involved in the process?
3. What three signal words helped you identify the sequence in this paragraph?

R23

Summarize

At the end of every lesson in this science book, you are asked to summarize. When you summarize, you use your own words to tell what something is about. In the lesson, you will find ideas for writing your summary. You can also use a graphic organizer like this one to summarize.

| Main Idea: Tell about the most important information you have read. | + | Details: Add details that answer important questions such as *who, what, where, when, why,* and *how*. | = | Summary: Retell what you have just read, including only the most important details. |

Tips for Summarizing

- To write a summary, first ask *What is the most important idea of the paragraph?*
- To add details, ask *who, what, when, where, why,* and *how.*
- Remember to use fewer words than the original.
- Tell the information in your own words.

Here and on the next page is an example.

> The water cycle is the constant recycling of water. As the sun warms the ocean, water particles leave the water and enter the air as water vapor. This is evaporation, the process of a liquid changing to a gas. Clouds form when water vapor condenses high in the atmosphere. Condensation occurs when the water vapor rises, cools, and changes from a gas to liquid. When the drops of water are too large to stay up in the air, precipitation occurs.

(Main Idea: first sentence; Details: remainder of paragraph)

Reading in Science Handbook

Here is what you could record in the graphic organizer.

Main Idea:		Details:		Summary:
The water cycle is the constant recycling of water.	+	Evaporation is the change from a liquid to a gas. Condensation is the change from a gas to a liquid. Precipitation is water that falls to Earth.	=	The constant recycling of water is the water cycle. It includes evaporation, condensation, and precipitation.

More About Summarizing

Sometimes a paragraph has details that are not important enough to be included in a summary. The graphic organizer remains the same because those details are not important to understanding the paragraph's main idea.

Skill Practice

Read the following paragraph. Use the Tips for Summarizing to answer the questions.

> Tides are the changes in the ocean's water level each day. At high tide, much of the beach is covered with water. At low tide, waves break farther away from the shore and less of the beach is under water. Every day most shorelines have two high tides and two low tides. High tides and low tides occur at regular times and are usually a little more than 6 hours apart.

1. If a friend asked you what this paragraph was about, what information would you include? What would you leave out?
2. What is the main idea of the paragraph?
3. Which two details would you include in a summary of the paragraph?

Draw Conclusions

At the end of each lesson in this science book, you are asked to draw conclusions. To draw conclusions, use the information that you have read and what you already know. Drawing conclusions can help you understand what you read. You can use a graphic organizer like this.

What I Read		What I Know		Conclusion:
Use facts from the text to help you understand.	+	Use your own experience to help you understand.	=	Combine facts and details in the text with personal knowledge or experience.

Tips for Drawing Conclusions

- To draw conclusions, first ask *What information from the text do I need to think about?*

- Then ask *What do I know from my own experience that could help me draw a conclusion?*

- Ask yourself whether the conclusion you have drawn is valid, or makes sense.

Here is an example.

Text information: Plants need air, nutrients, water, and light to live. A plant makes its own food by a process called photosynthesis. Photosynthesis takes place in the plant's leaves. In an experiment, a plant is placed in a dark room without any light. It is watered every day.

Here is what you could record in the graphic organizer.

What I Read		What I Know		Conclusion:
A plant needs air, nutrients, water, and light to live.	+	Plants use light to make the food they need to live and grow.	=	The plant will die since it is not getting any light.

Reading in Science Handbook

More About Drawing Conclusions

Sensible conclusions based on your experience and the facts you read are valid. For example, suppose the paragraph on page R26 included a sentence that said *After a day, the plant is removed from the dark room and placed in the sunlight.* You could then draw a different conclusion about the life of the plant.

What I Read		What I Know		Conclusion:
A plant needs air, nutrients, water, and light to live.	+	Plants use light to make the food they need to live and grow.	=	The plant will live.

Sometimes a paragraph might not contain enough information to draw a valid conclusion. Read the following paragraph. Think of one valid conclusion you could draw. Then think of one conclusion that would be invalid or wouldn't make sense.

> Cacti are plants that are found in the desert. Sometimes it does not rain in the desert for months or even years. Cacti have thick stems. The roots of cactus plants grow just below the surface of the ground.

Skill Practice

Read the following paragraph. Use the Tips for Drawing Conclusions to answer the questions.

> Animals behave in ways that help them meet their needs. Some animal behaviors are instincts, and some are learned. Tiger cubs learn to hunt by watching their mothers hunt and by playing with other tiger cubs. They are not born knowing exactly how to hunt.

1. What conclusion can you draw about a tiger cub that is separated from its mother?
2. What information from your own experience helped you draw the conclusion?
3. What text information did you use to draw the conclusion?

Using Tables, Charts, and Graphs

As you do investigations in science, you collect, organize, display, and interpret data. Tables, charts, and graphs are good ways to organize and display data so that others can understand and interpret your data.

The tables, charts, and graphs in this Handbook will help you read and understand data. The Handbook will also help you choose the best ways to display data so that you can draw conclusions and make predictions.

Reading a Table

A scientist is studying the rainfall in Bangladesh. She wants to know when the monsoon season is, or the months in which the area receives the greatest amounts of rainfall. The table shows the data she has collected.

Monthly Rainfall in Chittagong, Bangladesh	
Month	Rainfall (inches)
January	1
February	2
March	3
April	6
May	10
June	21
July	23
August	10
September	13
October	7
November	2
December	1

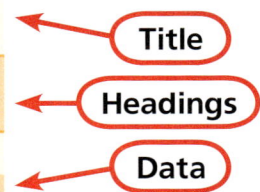

How to Read a Table

1. **Read the title** to find out what the table is about.
2. **Read the headings** to find out what information is given.
3. **Study** the data. Look for patterns.
4. **Draw conclusions.** If you display the data in a graph, you might be able to see patterns easily.

By studying the table, you can see how much rain fell during each month. If the scientist wanted to look for patterns, she might display the data in a graph.

R28

Reading a Bar Graph

The data in this bar graph is the same as that in the table. A bar graph can be used to compare the data about different events or groups.

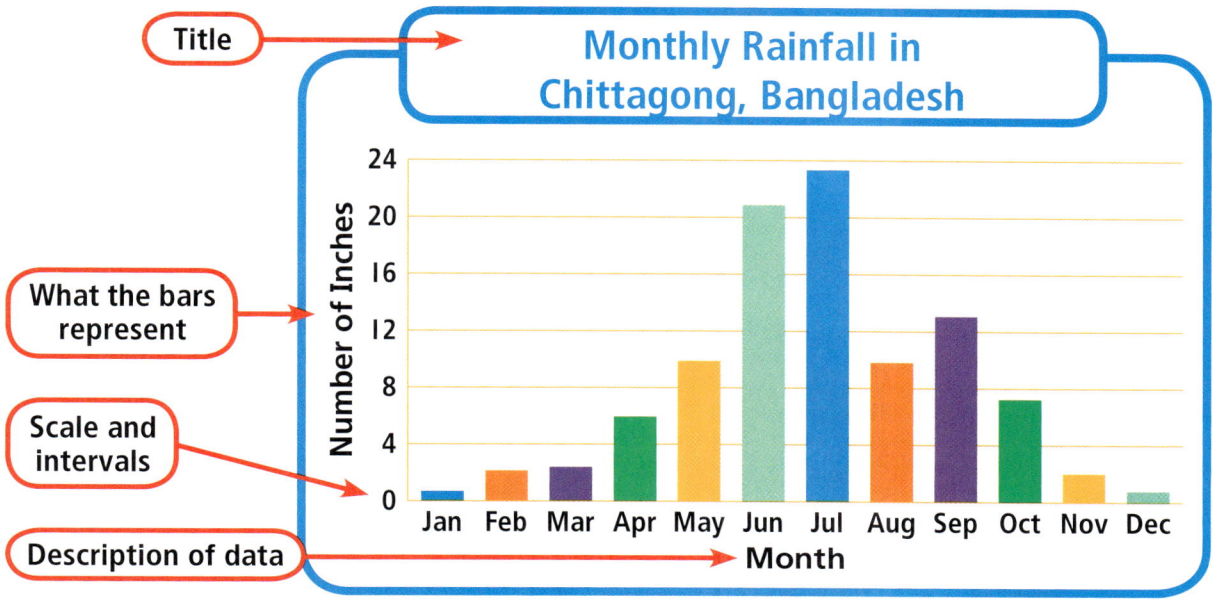

How to Read a Bar Graph

1. **Look** at the graph to determine what kind of graph it is.
2. **Read** the graph. Use the numbers and labels to guide you.
3. **Analyze** the data. Study the bars to compare the measurements. Look for patterns.
4. **Draw conclusions.** Ask yourself questions like the ones under Skills Practice.

Skills Practice

1. In which two months does Chittagong receive the most rainfall?
2. Which months have the same amounts of rainfall?
3. **Predict** During which months are the roads likely to be flooded?
4. How does the bar graph help you identify the monsoon season and the rainfall amounts?
5. Was the bar graph a good choice for displaying this data?

Reading a Line Graph

A scientist collected this data about temperatures in Pittsburgh, Pennsylvania.

Average Temperatures in Pittsburgh	
Month	Temperature (degrees Fahrenheit)
January	28
February	29
March	39
April	50
May	60
June	68
July	74
August	72
September	63
October	52
November	43
December	32

How to Read a Line Graph

1. **Look** at the graph to determine what kind of graph it is.
2. **Read** the graph. Use the numbers and labels to guide you.
3. **Analyze** the data. Study the points along the lines. Look for patterns.
4. **Draw conclusions.** Ask yourself questions like the ones under Skills Practice.

Here is the same data displayed in a line graph. A line graph is used to show changes over time.

- Title
- What the points represent
- Scale and intervals
- Description of data

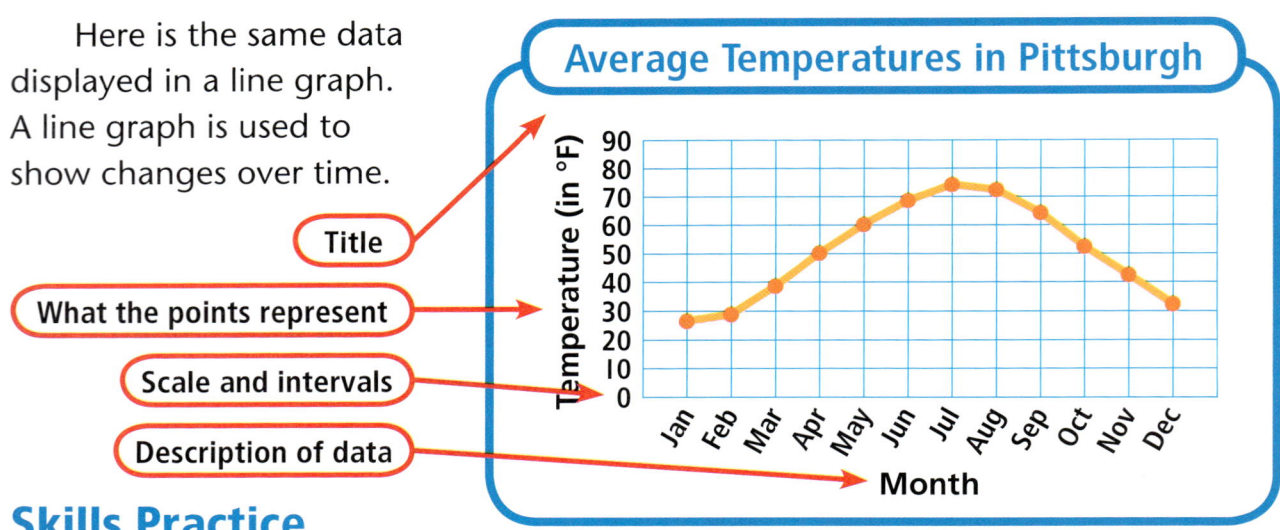

Skills Practice

1. In which three months are the temperatures the warmest in Pittsburgh?

2. **Predict** During which months are ponds in Pittsburgh likely to freeze?

3. Was the line graph a good choice for displaying this data? Explain why.

Math in Science Handbook

Reading a Circle Graph

Some scientists counted 100 animals at a park. The scientists wanted to know which animal group had the most animals. They classified the animals by making a table. Here is their data.

Animal Groups at the Park

Animal Group	Number Observed
Mammals	7
Insects	63
Birds	22
Reptiles	5
Amphibians	3

The circle graph shows the same data as the table. A circle graph can be used to show data as a whole made up of parts.

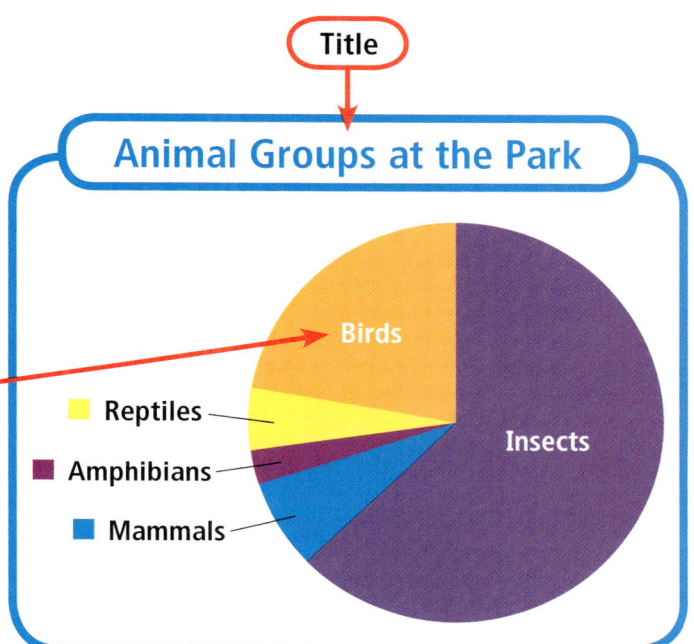

How to Read a Circle Graph

1. **Look** at the title of the graph to learn what kind of information is shown.

2. **Read** the graph. Look at the label of each section to find out what information is shown.

3. **Analyze** the data. Compare the sizes of the sections to determine how they are related.

4. **Draw conclusions.** Ask yourself questions like the ones under Skills Practice.

Skills Practice

1. Which animal group had the most members? Which one had the fewest?

2. **Predict** If you visited a nearby park, would you expect to see more reptiles or more insects?

3. Was the circle graph a good choice for displaying this data? Explain.

R31

Measurements

When you measure, you compare an object to a standard unit of measure. Scientists almost always use the units of the metric system.

Measuring Length and Capacity in Metric Units

When you measure length, you find the distance between two points. The table shows the metric units of **length** and how they are related.

Equivalent Measures
1 centimeter (cm) = 10 millimeters (mm)
1 decimeter (dm) = 10 centimeters (cm)
1 meter (m) = 1000 millimeters
1 meter = 10 decimeters
1 kilometer (km) = 1000 meters

You can use these comparisons to help you learn the size of each metric unit of length:

A **millimeter (mm)** is about the thickness of a dime.	A **centimeter (cm)** is about the width of your index finger.	A **decimeter (dm)** is about the width of an adult's hand.	A **meter (m)** is about the width of a door.
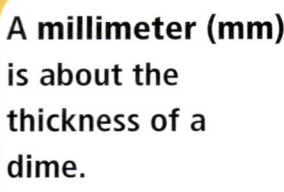	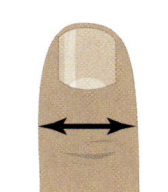	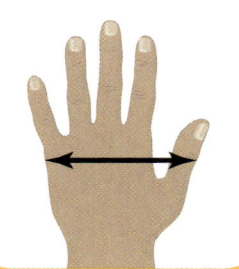	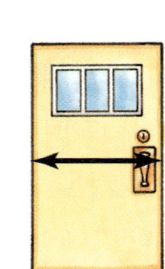

The following diagram shows how to multiply and divide to change to larger and smaller units.

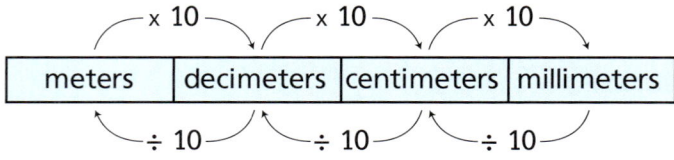

R32

Math in Science Handbook

When you measure capacity, you find the amount a container can hold when it is filled. The images show the metric units of **capacity** and how they are related.

A **milliliter (mL)** is the amount of liquid that can fill one part of a medicine dropper.

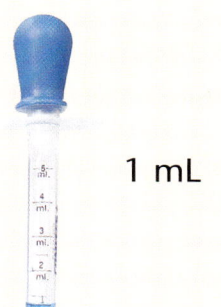

1 mL

A **liter (L)** is the amount of liquid that can fill a plastic water bottle.

1 L = 1000 mL

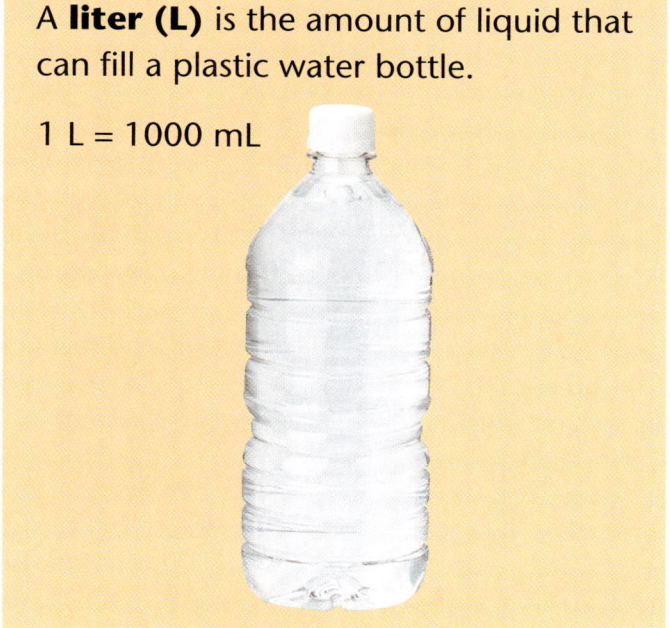

You can use multiplication to change liters to milliliters.

You can use division to change milliliters to liters.

2 L = ____ mL
Think: There are 1000 mL in 1 L.
2 L = 2 x 1000 = 2000 mL
So, 2 L = 2000 mL.

4000 mL = ____ L
Think: There are 1000 mL in 1 L.
4000 ÷ 1000 = 4
So, 4000 mL = 4 L.

Skills Practice

Complete. Tell whether you multiply or divide.

1. 3 L = ____ mL
2. 5000 mL = ____ L
3. 7000 mL = ____ L
4. 6 L = ____ mL
5. 500 dm = ____ cm
6. 4 m = ____ mm
7. 8 ____ = 80 cm
8. ____ m = 1400 cm

R33

Measuring Mass

Matter is what all objects are made of. **Mass** is the amount of matter that is in an object. The metric units of mass are the gram (g) and the kilogram (kg). You can use these comparisons to help you understand the masses of some everyday objects:

A paper clip is about **1 gram (g)**. 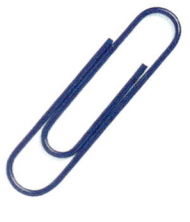	A slice of wheat bread is about **20 grams**. 	A box of 12 crayons is about **100 grams**. 	A large wedge of cheese is about **1 kilogram (kg)**. 

You can use multiplication to change kilograms to grams.

You can use division to change grams to kilograms.

2 kg = ____ g	4000 g = ____ kg
Think: There are 1000 g in 1 kg.	Think: There are 1000 g in 1 kg.
2 kg = 2 × 1000 = 2000 g	4000 ÷ 1000 = 4
So, 2 kg = 2000 g.	So, 4000 g = 4 kg.

Skills Practice

Complete. Tell whether you multiply or divide by 1000.

1. 5000 g = ____ kg
2. 3000 g = ____ kg
3. 4 kg = ____ g
4. 7 kg = ____ g

Math in Science Handbook

Measurement Systems

SI Measures (Metric)

Temperature
Ice melts at 0 degrees Celsius (°C).
Water freezes at 0°C.
Water boils at 100°C.

Length and Distance
1000 meters (m) =
 1 kilometer (km)
100 centimeters (cm) = 1 m
10 millimeters (mm) = 1 cm

Force
1 newton (N) = 1 kilogram x
 1 meter/second/second (kg-m/s^2)

Volume
1 cubic meter (m^3) =
 1 m x 1 m x 1 m
1 cubic centimeter (cm^3) =
 1 cm x 1 cm x 1 cm
1 liter (L) = 1000 millimeters (mL)
1 cm^3 = 1 mL

Area
1 square kilometer (km^2) =
 1 km x 1 km
1 hectare = 10,000 m^2

Mass
1000 grams (g) = 1 kilogram (kg)
1000 milligrams (mg) = 1 g
1000 kilograms = 1 metric ton

Rates
km/hr = kilometers per hour
m/sec = meters per second

Customary Measures

Temperature
Ice melts at 32 degrees
 Fahrenheit (°F).
Water freezes at 32°F.
Water boils at 212°F.

Length and Distance
12 inches (in.) = 1 foot (ft)
3 ft = 1 yard (yd)
5280 ft = 1 mile (mi)

Force
16 ounces (oz) = 1 pound (lb)
2000 pounds = 1 ton (T)

Volume of Fluids
2 cups (c) = 1 pint (pt)
2 pt = 1 quart (qt)
4 qt = 1 gallon (gal)

Area
1 square mile (mi^2) = 1 mi x 1 mi
1 acre = 4840 sq ft

Rates
mph = miles per hour
ft/sec = feet per second

Safety in Science

Doing investigations in science can be fun, but you need to be sure you do them safely. Here are some rules to follow.

1. **Think ahead.** Study the steps of the investigation so you know what to expect. If you have any questions, ask your teacher. Be sure you understand any caution statements or safety reminders.

2. **Be neat.** Keep your work area clean. If you have long hair, pull it back so it doesn't get in the way. Roll or push up long sleeves to keep them away from your activity.

3. **Oops!** If you should spill or break something, or get cut, tell your teacher right away.

4. **Watch your eyes.** Wear safety goggles anytime you are directed to do so. If you get anything in your eyes, tell your teacher right away.

5. **Yuck!** Never eat or drink anything during a science activity.

6. **Don't get shocked.** Be especially careful if an electric appliance is used. Be sure that electric cords are in a safe place where you can't trip over them. Don't ever pull a plug out of an outlet by pulling on the cord.

7. **Keep it clean.** Always clean up when you have finished. Put everything away and wipe your work area. Wash your hands.

Visit the Multimedia Science Glossary to see illustrations of these terms and to hear them pronounced.
www.hspscience.com

Glossary

As you read your science book, you will notice that new or unfamiliar terms have been respelled to help you pronounce them while you are reading. Those respellings are called *phonetic respellings*. In this Glossary you will see the same kind of respellings.

In phonetic respellings, syllables are separated by a bullet (•). Small uppercase letters show stressed syllables.

The boldfaced letters in the examples in the Pronunciation Key below show which letters and combinations of letters are pronounced in the respellings.

The page number (in parentheses) at the end of a definition tells you where to find the term, defined in context, in your book. Depending on the context in which it is used, a term may have more than one definition.

Pronunciation Key

Sound	As in	Phonetic Respelling	Sound	As in	Phonetic Respelling
a	bat	(BAT)	oh	over	(OH•ver)
ah	lock	(LAHK)	oo	pool	(POOL)
air	rare	(RAIR)	ow	out	(OWT)
ar	argue	(AR•gyoo)	oy	foil	(FOYL)
aw	law	(LAW)	s	cell	(SEL)
ay	face	(FAYS)		sit	(SIT)
ch	chapel	(CHAP•uhl)	sh	sheep	(SHEEP)
e	test	(TEST)	th	that	(THAT)
	metric	(MEH•trik)		thin	(THIN)
ee	eat	(EET)	u	pull	(PUL)
	feet	(FEET)	uh	medal	(MED•uhl)
	ski	(SKEE)		talent	(TAL•uhnt)
er	paper	(PAY•per)		pencil	(PEN•suhl)
	fern	(FERN)		onion	(UHN•yuhn)
eye	idea	(eye•DEE•uh)		playful	(PLAY•fuhl)
i	bit	(BIT)		dull	(DUHL)
ing	going	(GOH•ing)	y	yes	(YES)
k	card	(KARD)		ripe	(RYP)
	kite	(KYT)	z	bags	(BAGZ)
ngk	bank	(BANGK)	zh	treasure	(TREZH•er)

R37

Multimedia Science Glossary: www.hspscience.com

abiotic [ay•by•AHT•ik] Describes a nonliving part of an ecosystem (142)

absorption [ab•ZAWRP•shuhn] The taking in of light or sound energy by an object (427)

acceleration [ak•sel•er•AY•shuhn] Any change in the speed or direction of an object's motion (531)

adaptation [ad•uhp•TAY•shuhn] A body part or behavior that helps an organism survive (100)

air mass [AIR MAS] A large body of air that has a similar temperature and moisture level (290)

amplitude [AM•pluh•tood] A measure of the amount of energy in a wave (417)

anemometer [an•uh•MAHM•uht•er] A weather instrument that measures wind speed (296)

atom [AT•uhm] The smallest unit of an element that has all the properties of that element (376)

axis [AK•sis] The imaginary line around which Earth spins as it rotates (308)

bacteria [bak•TIR•ee•uh] Members of the kingdom of one-celled living things that lack nuclei (33)

barometer [buh•RAHM•uh•ter] A weather instrument used to measure air pressure (296)

basic needs [BAY•sik NEEDZ] Food, water, air, and shelter that an organism needs to survive (98)

bedrock [BED•rahk] The solid rock that forms Earth's surface (217)

biotic [by•AHT•ik] Describes a living part of an ecosystem (140)

carnivore [KAHR•nuh•vawr] An animal that eats only other animals (168)

change of state [CHAYNJ uhv STAYT] A physical change that occurs when matter changes from one state to another, such as from a liquid to a gas (384)

chemical change [KEM•ih•kuhl CHAYNJ] A reaction or change in a substance, produced by chemical means, that results in a different substance (394)

chemical energy [KEM•ih•kuhl EN•er•jee] Energy that can be released by a chemical reaction (508)

chemical property [KEM•ih•kuhl PRAHP•er•tee] A property that involves how a substance interacts with other substances (393)

chemical reaction [KEM•ih•kuhl ree•AK•shuhn] A chemical change (394)

clay [KLAY] The smallest particles that make up soil (218)

cold front [KOHLD FRUHNT] The boundary where a cold air mass moves under a warm air mass (292)

comet [KAHM•it] A ball of rock, ice, and frozen gases in space (320)

community [kuh•MYOO•nuh•tee] All the populations of organisms living together in an environment (136)

compound [KAHM•pownd] A substance made of two or more different elements that have combined chemically (394)

R38

Glossary

Multimedia Science Glossary: www.hspscience.com

condensation [kahn•duhn•SAY•shuhn] The process by which a gas changes into a liquid (271)

conduction [kuhn•DUK•shuhn] The movement of heat between two materials that are touching (450)

conductor [kuhn•DUK•ter] Materials that let electric charges travel through them easily (485)

constellation [kahn•stuh•LAY•shuhn] A pattern of stars that form an imaginary picture or design in the sky (326)

consumer [kuhn•SOOM•er] A living thing that can't make its own food and must eat other living things (166)

convection [kuhn•VEK•shuhn] The movement of heat in liquids and gases from a warmer area to a cooler area (451)

current electricity [KER•uhnt ee•lek•TRIS•uh•tee] A steady movement of charges through certain materials (476)

decomposer [dee•kuhm•POHZ•er] A living thing that feeds on the wastes and remains of plants and animals (170)

density [DEN•suh•tee] The amount of matter in an object compared to the space it takes up (344)

deposition [dep•uh•ZISH•uhn] The dropping of bits of rock and soil by a river as it flows (244)

direct development [duh•REKT dih•VEL•uhp•muhnt] A kind of growth in which an organism gets larger but doesn't go through other changes (84)

diversity [duh•VER•suh•tee] A great variety of living things (146)

earthquake [ERTH•kwayk] The shaking of Earth's surface caused by movement of rock in the crust (242)

ecosystem [EE•koh•sis•tuhm] A community and its physical environment together (132)

electric motor [uh•LEK•trik MOHT•er] A device that changes electrical energy to energy of motion (490)

electromagnet [ee•lek•troh•MAG•nit] A temporary magnet caused by an electric current (490)

element [EL•uh•muhnt] A substance made up of only one kind of atom (378)

energy pyramid [EN•er•jee PIR•uh•mid] A diagram showing how much energy is passed from one organism to the next in a food chain (180)

energy transfer [EN•er•jee TRANS•fer] A change of energy from one form to another (457)

environment [en•VY•ruhn•muhnt] All of the living and nonliving things surrounding an organism. (132)

erosion [uh•ROH•zhuhn] The process of moving sediment from one place to another (212)

evaporation [ee•vap•uh•RAY•shuhn] The process by which a liquid changes into a gas (270)

experiment [ek•SPAIR•uh•muhnt] A test of a hypothesis (15)

R39

Multimedia Science Glossary: www.hspscience.com

extinction [ek•STINGK•shuhn] The death of all the members of a certain group of organisms (118)

food chain [FOOD CHAYN] A series of organisms that depend on one another for food (176)

food web [FOOD WEB] A group of food chains that overlap (178)

force [FAWRS] A pull or push of any kind (532)

fossil [FAHS•uhl] The remains or traces of a plant or an animal that lived long ago (114, 250)

fossil record [FAHS•uhl REK•erd] The information about Earth's history that is contained in fossils (252)

frequency [FREE•kwuhn•see] A measure of the number of waves that pass in a second (417)

friction [FRIK•shuhn] A force that resists motion between objects that are touching (541)

fulcrum [FUL•kruhm] The fixed point on a lever (556)

fungi [FUHN•jy] Organisms that can't make food and can't move about (46)

galaxy [GAL•uhk•see] A huge system of many stars, gases, and dust (326)

gas [GAS] The state of matter that does not have a definite shape or volume (351)

gene [JEEN] The basic unit of heredity (67)

generator [JEN•er•ayt•er] A device that produces an electric current (492)

geothermal energy [jee•oh•THER•muhl EN•er•jee] Heat that comes from the inside of Earth (501)

glacier [GLAY•sher] A large, moving mass of ice (245)

gravitation [grav•ih•TAY•shuhn] A force that acts between any two objects and pulls them together (539)

gravity [GRAV•ih•tee] The force of attraction between Earth and other objects, the expression of gravitation (539)

habitat [HAB•ih•tat] An environment that meets the needs of an organism (174)

habitat restoration [HAB•ih•tat res•tuh•RAY•shuhn] Returning a natural environment to its original condition (153)

hail [HAYL] Round pieces of ice formed when frozen rain is coated with water and refreezes (277)

herbivore [HER•buh•vawr] An animal that eats only plants, or producers (168)

heredity [huh•RED•ih•tee] The process by which traits are passed from parents to offspring (66)

hibernation [hy•ber•NAY•shuhn] A dormant, inactive state in which normal body activities slow (107)

horizon [huh•RY•zuhn] A layer in the soil (217)

Glossary

Multimedia Science Glossary: www.hspscience.com

humus [HYOO•muhs] The remains of decayed plants or animals in the soil **(216)**

hurricane [HER•ih•kayn] A large tropical storm with wind speeds of at least 74 miles per hour **(278)**

hydroelectric power [hy•droh•ee•LEK•trik POW•er] Electrical energy made by using the kinetic energy of falling water **(500)**

hypothesis [hy•PAHTH•uh•sis] A statement of what you think will happen and why **(15)**

igneous rock [IG•nee•uhs RAHK] A type of rock that forms from melted rock that cools and hardens **(196)**

inclined plane [in•KLYND PLAYN] A simple machine that is a slanted surface **(570)**

inertia [in•ER•shuh] The property of matter that keeps an object at rest or keeps it moving in a straight line **(534)**

inference [IN•fer•uhns] An untested conclusion based on your observations **(12)**

instinct [IN•stinkt] A behavior that an animal begins life with and that helps it meet its needs **(106)**

insulator [IN•suh•layt•er] A material that does not let current electricity move through it easily **(480)**

intensity [in•TEN•suh•tee] A measure of how loud or soft a sound is **(411)**

invertebrates [in•VER•tuh•brits] The group of animals without backbones **(52)**

kinetic energy [kih•NET•ik EN•er•jee] The energy of motion **(499)**

land breeze [LAND BREEZ] A breeze that moves from the land to the water **(284)**

landform [LAND•fawrm] A natural feature on Earth's surface **(232)**

learned behavior [LERND bee•HAYV•yer] A behavior that an organism doesn't begin life with **(110)**

lever [LEV•er] A simple machine made of a bar that pivots on a fixed point **(556)**

life cycle [LYF CY•kuhl] All the stages a living thing goes through **(74)**

light [LYT] A form of energy that can travel through space and lies partly within the visible range **(440)**

liquid [LIK•wid] The state of matter that has a definite volume but no definite shape **(350)**

magnet [MAG•nit] An object that attracts iron and a few other (but not all) metals **(486)**

magnetic field [mag•NET•ik FEELD] The space around a magnet in which the force of the magnet acts **(489)**

magnetic poles [mag•NET•ik POHLZ] The parts of a magnet at which its force is strongest **(488)**

mass [MAS] The amount of matter in an object **(343)**

R41

Multimedia Science Glossary: www.hspscience.com

matter [MAT•er] Anything that has mass and takes up space (342)

mechanical energy [muh•KAN•ih•kuhl EN•er•jee] The total potential and kinetic energy of an object (509)

metamorphic rock [met•uh•MAWR•fik RAHK] A type of rock that forms when heat or pressure change an existing rock (198)

metamorphosis [met•uh•MAWR•fuh•sis] Major changes in the body form of an animal during its life cycle (86)

microscope [MY•kruh•skohp] A tool that makes an object look several times bigger than it is (6)

microscopic [my•kruh•SKAHP•ik] Too small to be seen with the eyes alone (33)

migration [my•GRAY•shuhn] The movement of animals from one region to another and back (108)

mineral [MIN•er•uhl] A solid nonliving substance that occurs naturally in rocks or in the ground (194)

mixture [MIKS•cher] A blending of two types of matter that are not chemically combined (358)

moon [MOON] A natural body that revolves around a planet (310)

motion [MOH•shuhn] A change of position of an object (522)

mountain [MOUNT•uhn] An area that is much higher than the land around it (232)

niche [NICH] The role of an organism in its habitat (175)

nonvascular [nahn•VAS•kyuh•ler] Without tubes or channels (44)

observation [ahb•zer•VAY•shuhn] Information from your senses (12)

omnivore [AHM•nih•vawr] An animal that eats both plants and other animals (168)

orbit [AWR•bit] The path of one object in space around another object (308)

organism [AWR•guh•niz•uhm] A living thing (32)

pan balance [PAN BAL•uhns] A tool that measures mass (8)

parallel circuit [PAIR•uh•lel SER•kit] A circuit that has more than one path for an electric current to follow (478)

phases [FAYZ•uhz] The different shapes that Earth's moon seems to have (310)

physical change [FIZ•ih•kuhl CHAYNJ] A change in matter from one form to another that doesn't result in a different substance (386)

physical property [FIZ•ih•kuhl PRAHP•er•tee] A trait that involves a substance by itself (393)

pitch [PICH] A measure of how high or low a sound is (410)

planet [PLAN•it] A large object that moves around a star (316)

pollution [puh•LOO•shuhn] Waste products that damage an ecosystem (152)

Glossary

Multimedia Science Glossary: www.hspscience.com

population [pahp•yuh•LAY•shuhn] All the individuals of the same kind living in the same ecosystem **(134)**

position [puh•ZISH•uhn] The location of an object **(522)**

potential energy [poh•TEN•shuhl EN•er•jee] Energy that an object has because of its position or its condition **(499)**

precipitation [pree•sip•uh•TAY•shuhn] Water that falls to Earth **(268)**

predator [PRED•uh•ter] A consumer that eats prey **(176)**

prey [PRAY] Consumers that are eaten by predators **(176)**

producer [pruh•DOOS•er] A living thing, such as a plant, that can make its own food **(166)**

protist [PROHT•ist] One of the kingdoms of living things that are one-celled **(33)**

pulley [PUHL•ee] A simple machine made of a wheel with a line around it **(562)**

radiation [ray•dee•AY•shuhn] The movement of heat without matter to carry it **(452)**

rain [RAYN] Precipitation that is liquid water **(276)**

rain shadow [RAYN SHAD•oh] The area on the side of a mountain range that gets little or no rain or cloud cover **(286)**

reflection [rih•FLEK•shuhn] The bouncing of light, sound, or heat off an object **(426, 441)**

refraction [rih•FRAK•shuhn] The bending of light when it moves from one kind of matter to another **(443)**

rock [RAHK] A solid substance made of one or more minerals **(194)**

rock cycle [RAHK SY•kuhl] The sequence of processes that change rocks from one type to another over long periods **(202)**

sand [SAND] The largest particles that make up soil **(218)**

scientific method [sy•uhn•TIF•ik METH•uhd] A way that scientists find out how things work and affect each other **(20)**

screw [SKROO] A simple machine made of a post with an inclined plane wrapped around it **(572)**

sea breeze [SEE BREEZ] A breeze that moves from the water to the land **(284)**

sedimentary rock [sed•uh•MEN•ter•ee RAHK] A type of rock that forms when layers of sediment are pressed together **(197)**

series circuit [SIR•eez SER•kit] A circuit that has only one path for an electric current to follow **(478)**

simple machine [SIM•puhl muh•SHEEN] A machine with few or no moving parts that you apply just one force to **(555)**

sleet [SLEET] Precipitation made when rain falls through freezing-cold air and turns to ice **(276)**

snow [SNOH] Precipitation made when water vapor turns directly into ice and forms ice crystals **(277)**

Multimedia Science Glossary: www.hspscience.com

solar energy [SOH•ler EN•er•jee] The power of the sun (502)

solar system [SOH•ler SIS•tuhm] A group of objects in space that revolve around a central star (316)

solid [SAHL•id] The state of matter that has a definite shape and a definite volume (350)

solubility [sahl•yoo•BIL•uh•tee] A measure of how much of a material will dissolve in another material (361)

solution [suh•LOO•shuhn] A mixture in which two or more substances are mixed completely (360)

speed [SPEED] The measure of an object's change in position during a unit of time (524)

spring scale [SPRING SKAYL] A tool that measures forces, such as weight (8)

standard measure [STAN•derd MEZH•er] An accepted measurement (4)

star [STAR] A huge ball of superheated gases (324)

state of matter [STAYT uhv MAT•er] One of three forms (solid, liquid, and gas) that matter can exist in (350)

static electricity [STAT•ik ee•lek•TRIS•uh•tee] An electrical charge that builds up on an object (474)

sun [SUHN] The star at the center of our solar system (324)

suspension [suh•SPEN•shuhn] A kind of mixture in which particles of one ingredient are floating in another ingredient (362)

topography [tuh•PAHG•ruh•fee] The shape of landforms in an area (234)

tornado [tawr•NAY•doh] A fast-spinning spiral of wind that touches the ground (278)

trait [TRAYT] A characteristic that makes one organism different from another (66)

transmission [tranz•MISH•uhn] The passing of light or sound waves through a material (428)

universe [YOO•nuh•vers] Everything that exists in space (326)

vascular [VAS•kyuh•ler] Having tubes or channels (42)

velocity [vuh•LAHS•uh•tee] The measure of the speed and direction of motion of an object (530)

vertebrates [VER•tuh•brits] The group of animals with backbones (50)

vibration [vy•BRAY•shuhn] A quick back-and-forth motion (408)

volcano [vahl•KAY•noh] A mountain that forms as lava flows through a crack onto Earth's surface (242)

volume [VAHL•yoom] The amount of space an object takes up (344)

Multimedia Science Glossary: www.hspscience.com

warm front [WAWRM FRUHNT] The boundary where a warm air mass moves over a cold air mass **(292)**

waste heat [WAYST HEET] Heat that can't be used to do useful work **(460)**

water cycle [WAW•ter SY•kuhl] The movement of water from the surface of Earth into the air and back again **(268)**

wavelength [WAYV•length] The distance between a point on one wave and the identical point on the next wave **(417)**

weathering [WETH•er•ing] The breaking down of rocks on Earth's surface into smaller pieces **(208)**

wedge [WEJ] A simple machine made of two inclined planes placed back to back **(574)**

weight [WAYT] A measure of the gravitational force acting on an object **(540)**

wheel-and-axle [weel•and•AK•suhl] A simple machine made of a wheel and an axle that turn together **(564)**

work [WERK] The use of a force to move an object over a distance **(554)**

Index

Abdomen (insects), 54
Abiotic factors, 142
    climate, 144–145
    protection of, 154
Absorption
    definition of, 427
    of light, 442
    by nonvascular plants, 44
    of sound, 427
Acceleration
    definition of, 531
    force and, 532–533
    mass and, 534
    velocity and, 531
Acid rain, 152
Acquired traits, 69
Adaptations, 100–101
Adolescent stage (humans), 85
Adulthood stage (humans), 85
Agate, 194
Age of Mammals, 254
Air
    in ecosystems, 142
    mass of, 375
    moisture level in, 290
    in soil, 216
    sound carried by, 424
Air masses, 290–292
Air pollution, 152
Air pressure, 295
Ako, Japan, festival, 406
Aleutian Islands (Alaska), 236
Algae, 36, 38, 166
Alligators, 99, 177
Alpha Centauri, 325
Alpine glaciers, 245
Amanita mushrooms, 46
Amber, 251
Ammonites, 248
Amoebas, 36

Amphibians, 51
    metamorphosis of, 86
    reproduction of, 83
Amplitude, 417, 419
Andesite, 196, 204
Anechoic chambers, 427
Anemometers, 296
Anemones, 175
Animals
    adaptations of, 100, 101
    basic needs of, 98–99
    cells of, 34
    climate and, 144
    in ecosystems, 140, 141
    in Everglades, 136
    extinct, 118
    fossil vs. present-day, 114–115
    fossils of, 251
    growth and development of, 84–85
    hibernation of, 101, 107
    instinctual behaviors of, 106–108
    invertebrates, 52–53
    learned behaviors of, 110
    life cycles of, 82–86
    migration of, 108–109
    natural fertilizers from, 220
    populations of, 134
    prey and predators, 176–177
    producers vs. consumers, 166–167
    vertebrates, 50–51
    weathering and, 211
    See also Living things
Antarctic Ocean food web, 179
Antennae (insects), 54
Anvil (ear), 425, 428
Apollo, 306
Aquanauts, 156–157
***Aquarius* underwater laboratory,** 156–157

Arachnids, 52, 53
Archer fish, 164
Arctic tern, 109
Arthropods, 53
Artificial fertilizers, 220
Artificial muscle, 512–513
Asexual reproduction
    in plants, 78
Aspirin, 151
Asteroids, 316, 321
    extinction caused by, 118
Atacama Desert (Chile), 138
Atoms, 376–377
    of elements, 378
Autumn equinox, 309
Axis, Earth's, 308, 309
Axles. *See* Wheel-and-axles

Backbones, animals with. *See* Vertebrates
Bacteria, 33, 35–37
Balances, 8
Bald eagles, 84
Ballard, Michael, 258
Ballard, Robert, 430–431
Banana slugs, 48
Bar graphs, R29
Barnard's Star, 325
Barometers, 296
Barred spiral galaxies, 326, 327
Barrier islands, 236, 246
Basalt, 196, 206, 219
Basic needs, 98–99
Bats, 100
    hibernation of, 107
Batteries, energy in, 508
Beaches
    formation of, 246
Beaks, bird, 96
Bears, 110
Beavers, 99

R46

Index

Bedrock, 217, 219
Bees
   metamorphosis, 86
Behaviors
   instinctual, 69, 106–108
   learned, 110
Bel (sound unit), 412
Bell, Alexander Graham, 412
Beluga whales, 104
Bicycles, 152, 541, 542
   glow-in-the-dark, 464
BioBlitz, 56–57
Biologists, 122
Biotic factors, 140, 154
Biotite, 195
Birds, 51
   adaptations of, 100, 101
   in Everglades, 136
   extinct, 118
   instincts of, 106
   migration of, 109
   reproduction of, 83
   sounds made by, 418
   talons of, 129
Blackbirds, 135
Blizzards, 278
Blood, R5
Blueberries, 398–399
Blue lupines, 158
Blue stars, 325
Boiling point, 354
Bones
   in ear, 425
   fossilized, 114
Boron, 484
Bracket fungi, 170
Bread molds, 46
Breezes, 284
Bristlecone pine, 83, 117
British Isles, 236
Bryophytes, 44, 45
Bubbles, 382
Bug box, 6
Bulbs (plant), 78

Bushway, Al, 398–399
Butterflies, 101
   metamorphosis, 86
Buttes, 234

C

Calcite, 194
Calendars, 312
Calories, 180
Camels, 115
Camera trapping, 182–183
Canopy (rain forests), 146
Canyons, 229, 233
Capacity, measuring, R33
Carbon, 378
Careers
   biologist, 122
   drafter, 578
   food manufacturer, 400
   landscaper, 224
   nuclear medicine technologist, 366
   paleontologist, 184
   power plant technician, 514
   zoo veterinarian, 90
Caribou, 108
Carnivores, 168, 169
Cars
   hybrid, 510
   pollution from, 152, 153
   waste heat from, 461
Casts, fossil, 250
Cats
   growth and development of, 84
Cause and effect, identifying, R20–R21
Cell membrane
   in one-celled organisms, 36
Cells, 34–35
   animal, 34
   functions of, 34, 35
   genes in, 67
   plant, 35

Cenozoic Era, 254
Ceramic insulators, 472
Chalcopyrite, 195
Change of state (matter), 384–385
   chemical, 394–396
   dissolving, 388
   physical, 386–387
Changing variables
   energy roller coaster, 497
Charge, electrical, 474
Chemical changes, 394–395
   recognizing, 396
Chemical energy, 508–509
Chemical properties of matter, 392–393
Chemical reactions, 390, 393, 394
Childhood stage (humans), 85
Chipmunks, 33, 177
Chlamydomonas, 38
Chlorophyll, 43
Chloroplasts, 35, 36, 43
Christmas Tree worm, 52
Chromatium, 38
Chrysalis, 86
Cinder cone volcanoes, 242
Circle graphs, R31
Circuits. *See* Electrical circuits
Circular motion, 536
Circulatory system, R4–R5
   in vertebrates, 50
Cirrus clouds, 293
Clams, 52, 250
Clark, Christopher Ray, 332
Classification, 13
   energy sources and uses, 505
   fungi, 46
   invertebrates, 52–53
   of living things, 32–38
   nonvascular plants, 44–45
   one-celled organisms, 33–38
   vascular plants, 42–43
   vertebrates, 50–51
Clay, 218
Cliffs, weathering of, 246

R47

Climate, 133, 144–145
    fossil record of changes in, 253
Clones (plants), 78
Closed circuits, 477
Clouds
    formation of, 268, 271
    types of, 293
    in water cycle, 269
Coal, 508
Coastal plains, 234
Cochlea, 425, 428
Coconut, 74
Coelancanth fish, 115
Cold air masses, 290–291
Cold fronts, 292, 294
Color, 346
    as animal adaptation, 100
    on weather maps, 294–295
Comet moths, 80
Comets, 320
Communication, 16
    food chain, 173
    path of light, 438
    in scientific method, 22
    walking locations, 521
Communities, 136
Comparison, 13
    animal life cycles, 81
    effects of water, 139
    electricity and magnetism, 485
    in reading, R18–R19
Compasses, 489, 494
Complete metamorphosis, 86
Composite volcanoes, 242
Compost, 220
Compound eyes, 54
Compounds, 394, 395
Computers, 461, 494
Concert halls, 405
Condensation, 271
Conduction (heat), 450
Conductors, electric, 480–481
Cone-bearing plants, 43
Conforti, Kathleen, 183
Conglomerate (rock), 197

Conservation
    of energy, 510
    of natural resources, 153
Constellations, 326, 328
Consumers, 166–167, 178, 180
Contour plowing, 148
Contrasting (reading), R18–R19
Control variables, 15
Convection, 451
Cooking heat, 458
Cooling, states of matter and, 352
Copper, 195
Coral islands, 236
Coral reef habitat, 130, 175
Corals, 52
Core (Earth), 240
Corn, 116
Crabs, 53, 175
Crustaceans, 53
Crust (Earth), 202, 240. *See also* Landforms
Cumbre Vieja volcano, 256–257
Cumulus clouds, 293
Curie, Marie, 366
Current electricity, 476–477
    in electromagnets, 490
Currents, water vs. electric, 476–477
Cyanobacteria, 37

D

Daffodils, 78
Dams, 500
Dandelion seeds, 72
Data, 16
Daum, Peter, 400
Da Vinci, Leonardo, 578
Davis, Peter, 544–545
Day, Simon, 256–257
Decibels (dB), 412
Decomposers, 170
Deer, 84
Deltas, 235, 244

Democritus, 376, 377
Density, 344–345
    of water, 340
Deposition, 244
Deserts
    climate of, 144, 145
    habitats of, 138, 174
    soils of, 219
Details, supporting, R16–R17
Diana, 306
Diatoms, 36, 38
Diet, R14
Digestive system, R2–R3
    in vertebrates, 50
Dinoflagellates, 36, 38
Dinosaurs, 250–254
Diorite, 196
Diplococci, 37
Direct development, 84
Direction
    force and, 532
    velocity and, 530
Diseases
    bacteria causing, 37
    human, R12–R13
Displaying data, 16
    densities of liquids, 341
Dissolving, 388
Distance
    speed and, 524–525
    tools for measuring, 4
Diversity (ecosystems), 146
Dogs, 64, 70, 110
    growth and development of, 84
Dog sleds, 532–533
Drafters, 578
Dragonflies, 135
Drawing conclusions, 16
    about eating like birds, 97
    about inherited characteristics, 65
    about moving up, 569
    in reading, R26–R27
    in scientific method, 22

Index

and solutions, 373
about wet steel wool, 391
Drinking water, 272
Droppers, 5
Duckweed, 40
Dunes, 235, 246

E

Eardrum, 425, 428
Ears, human, 424, 425, R11
Earth, 317
   axis of, 308, 309
   changes in crust, 202. *See also* Landforms
   force on sun, 538
   fossil record and history of, 252–253
   geologic time scale of, 254
   layers of, 240–241
   orbit of, 308, 309
   seasons and tilt of, 308–309
Earth Day, 300
Earthquakes, 243
Earthworms, 52
Eclipses, 332
Ecosystems, 132–133
   adaptations in, 100–101
   basic needs in, 98–99
   climate and, 144–145
   climate of, 133
   communities, 136
   coral, 130
   definition of, 132
   diversity in, 146
   growth and decay in, 102
   humans within, 150–154
   individuals in, 134
   living things affecting, 140–141
   natural resources in, 150
   negative/positive changes in, 152–153
   nonliving things in, 142–143
   planning for changes in, 154
   populations of, 134–135

Edison, Thomas, 506, 514
Eggs
   of animals, 82, 83
   dinosaur, 251, 253
   fern, 77
   and reproduction, insect, 86
Electrical circuits, 476–479
Electricity
   charges, 474, 538
   circuits, 478–479
   conductors, 480–481
   current, 476–477
   electromagnets, 490–491
   force between charges, 538
   as fuel, 458
   generators, 492–493
   insulators, 480–481
   motors, 493
   static, 474–475
   switches, 482
   uses of, 506–507
   from walking, 512–513
Electroluminescence, 464
Electromagnets, 490–491
Elements, 378–379
   definition of, 378
   metals vs. nonmetals, 380
Elephants, 84, 184, 252
Energy, 498–499
   chemical, 508–509
   conservation of, 510
   food chains, 176–177
   food webs, 178–179
   geothermal, 462–463, 501
   heat and, 449
   hydroelectric, 500
   kinetic, 499, 500
   measurement of, 504
   mechanical, 509
   potential, 488, 499, 500
   solar, 324, 456–457, 502
   thermal, 449. *See also* Heat
Energy pyramids, 180
Energy transfer, 457
Engines
   jet, 419

   steam, 460
Environment, 132
   fossil record of changes in, 253
   traits influenced by, 68
Equinoxes, 309
Eras, 254
Erosion, 152, 212
   contour plowing and, 148
   definition of, 212
   of soil, 152, 153
   by wind and waves, 246
Evaporation, 270, 388
Everglades National Park (Florida), 136
Exercise, R15
Exoskeletons, 53, 54
Experiments, 15
   making circular motion, 536
   in scientific method, 21
   sprouting seeds, 73
   straw models, 19
   testing soil, 215
   training fish, 105
Extinction, 118
   of dinosaurs, 254
Eyes, R10
   of insects, 54
   seeing light, 444

F

Faults, 243
Ferns, 33, 43, 76
   life cycles of, 77
Fertilizers, 220
Fiddleheads, 77
Fields, magnetic, 489
Fiords, 245
Firefighting, 364–365
First-level consumers, 178
FIRST program, 577
Fish, 51
   coelacanth, 115
   growth and development of, 84

R49

hibernation, 107
reproduction of, 83
as vertebrates, 50
Fishing rods, 566
Flies
fruit flies, 84
metamorphosis, 86
Floodplains, 234, 244
Floods, 274, 280
Flowering plants, 43
development of, 74–75
Flowering stage, 75
Fluorite, 194
Food, need for, 98
Food chains, 176–177
Food manufacturers, 400
Food webs, 178–179
Fool's gold, 195
Force(s)
acceleration and, 531–534
changing direction of, 562
definition of, 532
electric, 538
friction, 541–542
gravitation, 539
gravity, 539
with inclined planes, 570
with levers, 556
natural, 538
with simple machines, 554
with wedges, 574
with wheel-and-axle, 564
work and, 554
Forceps, 6
Forests
as ecosystem, 133
rain forest floor, 146
Fossey, Dian, 90
Fossil fuels, 508, 510
Fossil record, 252–253
Fossils, 250–251
ammonites, 248
animal, 114–115
bacteria in, 37
definition of, 114, 250

formation of, 250
plant, 116–117
Foxglove, 151
Frame of reference, 523
Freezing, 385
Freezing point, 354
Frequency
of sound, 418
of waves, 417
Friction, 541, 542
Frogs, 86, 134
hibernation, 107
Fronds, fern, 77
Fronts (weather), 292, 294
Fruit flies, 84
Fruiting stage, 75
Fuels, 458
fossil, 508
Fulcrum, 556
Full moon, 310, 311
Fungi, 33, 46

G

Gabbro, 196
Galápagos Island land tortoises, 83
Galaxies, 326
Galena, 195
Galileo, 314
Gametophyte generation, 77
Gas giants, 318, 319
Gases, 351, 384, 385
changes in state of, 354
condensation and, 271
mixtures of, 359
water vapor, 268
Gasoline, 152, 508
Gaspara, 320
Gathering data, 16
flood modeling, 275
Generators, 471, 492–493, 509
Genes, 67
trait variations from, 70
Geologic time scale, 254

Geothermal energy, 462–463, 501
Germination, 74, 75
Geysers Power Plant (California), 501
Giant's Causeway (Ireland), 206
Ginkgo tree, 117
Glaciers, 245
erosion and, 212
valley formation by, 233
Gliders, 578
Gneiss, 198
Gods, ancient, 306
Gold, 195, 380
Grafting, 78
Granite, 194–196, 219
Graphs, R29–R31
Grasshoppers, 53, 86
Grasslands
climate of, 144
as ecosystem, 132
soils of, 218, 219
Gravitation, 539
weight and, 540
Gravity, 539
Gray whales, 101, 108
Great auks, 118
Great Sphinx (Egypt), 222–223
Groundwater, 272
Guitar strings, 408, 414
Gypsum, 194, 235
Gypsy moths, 140
Gyroscopes, 576

H

Habitats, 174–175
extinction and changes in, 118
organisms in, 32
restoration of, 153
Hail, 268, 276, 277
Halley's comet, 321
Hammer (ear), 425, 428

Index

Hand lenses, 6
Handling, tools for, 6
Hand trucks, 557
Hawai'i, 236
Hawks, 129, 177
Head (insects), 54
Hearing, 424, R11
   process of, 425, 428
   sound waves and, 409
   See also Sound
Heat
   conduction of, 450
   convection of, 451
   motion from, 460
   radiation of, 452
   rock formation and, 198
   from solar energy, 502
   sources of, 456–457
   states of matter and, 352
   temperature and, 448–449
   use of, 459–460
   waste, 461
Heat pumps, 463
Hedge clippers, 558
Henry I, King of England, 4
Herbivores, 168, 169
Heredity, 66–67
   definition of, 66
   genes, 67
   traits, 68–69
   variations in, 70
Herons, 69, 99
Hibernation, 101, 107
Hills, 232
Himalayas, 241
Hine's emerald dragonfly, 135
Holzhauer, Madeline, 58
Horizons (soil), 217–219
Hornblende, 195
Hornwort, 45
Horses, 100, 110
Horseshoe crabs, 84
Hot springs, 462–463
Hubble Space Telescope, 322, 327

Human body
   blood, R5
   cells of. *See* Cells
   diet for, R14
   systems of. *See* Systems (human body)
Humans
   in ecosystems, 150–154
   life cycle of, 85
   population of, 154
Hummingbirds, 51
Humpback whales, 422
Humus, 216–220
Hurricane Hunters, 298–299
Hurricanes, 278–279, 298–299
Hybrid cars, 510
Hydrangea, 68
Hydroelectric power, 500
Hypothesizing, 15
   about drops of liquids, 383
   about ruler vibrations, 407
   in scientific method, 21

I

iBOT, 576
Ice
   erosion and, 212
   glaciers, 245
   weathering and, 210
Ice Age, 212, 252
Ice sheets, 245
Ice storms, 288
Identifying variables, 15
   hearing sounds, 423
Igneous rocks, 196, 204
Iguanas, 51
Illness, R12–R13
Inclined planes, 570–571
   screws, 572
   wedges, 574
Incomplete metamorphosis, 86
Individuals (in ecosystems), 134, 135
Inertia, 534

Infant stage
   animals, 82
   humans, 85
Inference, 12
   barometers, 289
   breakdown of chalk, 207
   erosion, 149
   making fossils, 113
   making telescopes, 323
   melting, boiling, evaporation, 349
   salt to fresh water, 267
   solar hot spots, 455
Inner core (Earth), 240
Inner planets, 316, 317
Inquiries
   skills for, 10–16
   tools for, 2–8
Insects, 53, 54
   life cycles of, 83
   metamorphosis, 86
   molting by, 84
Instincts, 69, 106
   hibernation, 107
   migration, 108
Insulators, 472, 480–481
Insulin, 122
Intensity (of sound), 411–412
International System (SI), 14, R35. *See also* Metric system
Interpreting data, 16
   flood modeling, 275
   moving up, 569
Invertebrates, 52–53
Investigation, 15
Iron, 379
   atoms of, 372
   in magnets, 484
   melting/boiling points of, 354
   sinkhole formation and, 209
Islands, 236
   volcanic, 236, 241

Jaguar reserve (Belize), 182–183
Jaguars, 169
Jellyfish, 52
Jet engines, 419
Jupiter, 318

Kamen, Dean, 576–577
Kangaroos, 50, 82, 83
Kansas, fossils in, 253
Kennedy, John F., 430
Kilowatt-hours, 504
Kinetic energy, 499, 500
Kingdoms, 33
    animal, 50
    plant, 42
Koi fish, 107

La Brea tar pits (California), 118
Land breezes, 284
Landforms
    canyons, 233
    deltas, 235
    dunes, 235
    earthquakes and, 243
    glaciers and, 245
    hills, 232
    islands, 236
    mountains, 232
    plains, 234
    plateaus, 234
    rivers and, 244
    valleys, 233
    volcanoes and, 241–243
    water cycle and, 284–286
    waves and, 246
    weathering and, 208
    wind and, 246

Landscapers, 224
Land tortoises, 83
Lasers, 437
Latimer, Lewis, 514
Lava, 196, 200
Lead, 195
Learned behaviors, 110
Leaves, 43
Leden, Judy, 578
Lee, Melinda, 546
Legs (insects), 54
Length, measuring, 4, R32
Levers, 555–558
Life cycles, 102
    of animals, 82–86
    definition of, 74
    of ferns, 77
    human, 85
    metamorphosis, 86
    of plants, 73–78
Life spans, 83
Light
    absorption of, 442
    definition of, 440
    path of, 440, 443
    properties of, 440
    reflection, 426, 441
    refraction, 443
    seeing, 444
Light bulbs, 476–477
    circuits in, 479
    conductors and insulators in, 480–481
    first, 506
    switches for, 482
Lightning, 286, 475
Limestone, 194, 219
Line graphs, R30
Lion fish, 175
Lions, 106
Liquids, 350–351, 384, 385
    changes in state of, 384
    condensation, 271
    density of, 344–345
    evaporation, 270

    measuring, 5
    mixtures of, 359
Liverwort, 45
Living things
    adaptations of, 100–101
    basic needs of, 98–99
    cells of, 34–35. *See also* Cells
    classification of. *See* Classification
    consumers, 166–167
    decomposers, 170
    in ecosystems, 132–133, 140–141
    energy pyramids, 180
    as environment, 132
    food chains, 176–177
    food webs, 178–179
    geologic time scale of, 254
    growth and decay of, 102
    habitats, 174–175
    producers, 166
    roles of, 164–170
    in soil, 216
    weathering and, 210, 211
    See also Animals; Human body; Plants
Longitudinal waves, 417
Loudness, 419
Lunar eclipse, 332
Lunar year, 312
Luster (minerals), 195

Machines, 555
    simple, 555. *See also* Simple machines
    working together, 566
Magma, 196, 198, 240, 241
Magnetic fields, 489, 492
Magnetic poles, 488–489
Magnetic resonance imaging (MRI), 491
Magnetism
    magnetic fields, 489, 492
    magnetic poles, 488–489

Index

Magnets, 484, 486–487
   definition of, 486
   electromagnets, 490–491
   uses of, 494
Magnifying boxes, 6
Magnifying lenses, 6
Main ideas (reading), R16–R17
Mammals, 51
   learned behaviors in, 110
   life cycles of, 82
Mammoths, 252
Mangrove swamps, 135, 176
Mantle (Earth), 240
Maps, weather, 294–295
Marble, 194, 198
Mars, 317, 330–331
Mars Rovers, 2, 18
Marsupials, 82–83
Mass, 343, 374–375
   acceleration and, 534
   measuring, 8, R34
   thermal energy and, 449
   weight vs., 540
Matter, 342, 346
   basic properties of, 374–375
   chemical changes in, 394–395
   chemical properties of, 392–393
   definition of, 342, 374
   density of, 344–345
   electrical charge of, 474
   elements, 378–379
   mass of, 343
   particles of, 376–377
   physical changes in, 386–387
   physical properties of, 346, 392–393
   states of. *See* States of matter
   volume of, 344
Measurement, 14
   of acceleration, 531
   of capacity, R33
   customary system of, R35
   decomposing bananas, 165
   of distance, 4
   of length, R32
   of mass, R34
   metric, R32–R35
   of motion, 524–525, 529
   seasons and sunlight, 307
   of sound intensity, 412
   with straws, 3
   systems of, R35
   tools for, 8
   of volume, 5
   of weather, 296
Measuring tape, 4
Mechanical energy, 509
Medicines, 151
Melting, 385
Mercury (element), 379
   melting/boiling points of, 354
Mercury (planet), 316, 317
Mesas, 234
Mesozoic Era, 254
Metals, 380
   changes in states of, 354
   as electrical conductors, 480
   magnets and, 484, 486
Metamorphic rocks, 198, 204
Metamorphism, 198
Metamorphosis, 86
Meter, 4
Metric system, 14, R32–R35
   capacity measurement, R33
   length measurement, R32
   mass measurement, R34
Microscopes, 6, 7
Microscopic organisms, 33. *See also* One-celled organisms
Migration, 101, 108–109
Milky Way, 326
Millipedes, 170
Mineral resources, 150
Minerals, 194–195
   definition of, 194
   in fossils, 250–251
   in rock, 194, 198
Mirrors
   reflections by, 438, 441
   solar energy and, 456
Mixtures, 358–359
   solutions, 360
   suspensions, 362
Modeling, 14
   cells, 31
   distances between planets, 315
   ecosystem, 131
   landforms, 231
   rock cycle, 201
   sedimentary rock, 193
   with straw, 11
   volcanic eruptions, 239
Moisture level (air), 290
Molds, fossil, 250
Mollusks, 52
Molting, 84
Monarch butterflies, 88–89, 101
Mono Lake (California), 356
Months, 312
Moon(s)
   calendars and, 312
   of inner planets, 317
   orbit of, 310, 311
   of outer planets, 318
   phases of, 310–311
Morels, 46
Mosquitoes, 136, 172, 432
Mosses, 43, 45, 76
Moths, 80, 86
Motion
   changing positions, 522–523
   definition of, 522
   force and acceleration, 532–533
   from heat, 460
   mass and acceleration, 534
   measuring, 524–525
   velocity, 530–531
Motors, electric, 493
Mount Saint Helens, 242, 258
Mountains, 232
   formation of, 198
   Himalayas, 241
   rain shadows, 286
   rock cycle and formation of, 204
   volcanoes, 242

MRI (magnetic resonance imaging), 491
Mudstone, 198
Mullets, 176
Multicelled organisms, 34
Munsell, Albert H., 224
Muscle, artificial, 512–513
Muscular system, R8–R9
  in vertebrates, 50
Mushrooms, 46
Musical instruments, 408, 410, 418
Mussels, 52

National Weather Service (NWS), 295
Natural fertilizers, 220
Natural forces, 538
Natural gas, 508
Natural resources
  in ecosystems, 150
  humans' uses of, 151
Neodymium, 484
Neptune, 318, 319
Nerves, hearing and, 425
Nervous system (vertebrates), 50
New moon, 310, 311
Niche, 175
Nickerson, Dorothy, 224
Niezrecki, Chris, 464
Nimbostratus clouds, 293
Nitrogen, melting/boiling points of, 354
Nonliving things (in ecosystems), 132, 142–143
Nonmetals, 380
Nonrenewable resources, 150
Nonvascular plants, 44–45
Northern Hemisphere
  constellations visible in, 328
  seasons in, 309
  solstices and equinoxes in, 309
Nuclear medicine technologists, 366
Nucleus (one-celled organisms), 36
Number skills. *See* Using numbers
Nurse logs, 102
Nurture, 68
Nutrients in soil, 220
Nutrition, R14
NWS (National Weather Service), 295
Nyberg, Michael, 432
Nymphs, 86

Observation, 12
  animal tracks, 249
  backbones, 49
  heating of land/water, 283
  landform models, 231
  making telescopes, 323
  plant stems, 41
  rubber bands, 553
  in scientific method, 20
  tools for, 6–7
Oceans
  food web in, 178–179
  fossil record and changes in, 252
  invertebrates in, 52
  new crust in, 241
  plates and, 241
  water in, 266
Offspring, inherited traits of, 66–67
Omega Nebula, 322
Omnivores, 168, 169
One-celled organisms, 33, 34, 36
  bacteria, 37
  in Precambrian Era, 254
  protists, 38
  sponges, 52
Opaque materials, 442
Orbits
  of Earth, 308, 309
  gravitation and, 539
  of moon, 310
Orb-weaver spiders, 106
Ordering, 81. *See also* Classification
Oregon coast, 238
Organisms
  classification of, 32–33
  definition of, 32
  multicelled, 34
  one-celled, 33, 34. *See also* One-celled organisms
  See also Living things
Organs, human. *See also specific organs, for example:* Stomach
Orion, 326, 328
Oscilloscope, 420
Outer core (Earth), 240
Outer ear, 425
Outer planets, 316, 318–319
Oxygen, 378
  melting/boiling points of, 354
Oyster mushrooms, 46

Pacific Ocean, coral islands in, 236
Pacific yew tree, 151
Pahoehoe lava, 200
Paleontologists, 184
Paleozoic Era, 254
Palo Duro Canyon (Texas) 233
Pan balances, 8
Paper cutter, 558
Paradise tree snakes, 120–121
Parallel circuits, 478–479
Parents, inherited traits from, 66–67
Particles (matter), 350–352, 376–377
Pathogens, R12–R13
Peninsulas, 285
Petrified Forest (Arizona), 251
Petrified wood, 116, 251
Phases (of moon), 310–311

Index

Phoenix Boulder Blast (Arizona), 560
Photosynthesis, 43
Physical changes, 386–387
   dissolving, 388
Physical properties, 392–393
   of minerals, 194
   of soil, 218
Phytoplankton, 179
Piersanti, Michaela, 300
Pilobolus, 46
Pine trees, 74
Pitch, 410, 418
Plains, 234
Planets, 316
   definition of, 316
   inner, 317
   outer, 318–319
   years on, 319
Planning investigations
   dissolving solids, 357
Plants
   asexual reproduction in, 78
   cells of, 35
   climate and, 144
   in ecosystems, 140, 141
   extinct, 118
   as food source, 98
   fossil vs. present-day, 116–117
   life cycles of, 74–78
   nonvascular, 44–45
   from seeds, 74–75
   soil and, 220
   from spores, 76–77
   vascular, 42–43
   See also Living things
Plateaus, 234
Plates, 240–241
   movement of, 241
   volcano formation and, 242
Platypus, 95
Pluto, 318, 319
Poles, magnetic, 488–489
Pollen, 75
Pollination, 75
Pollution, 152–153

Polyacrylate, 365
Populations (ecosystems), 134–135
   balance in, 141
   in communities, 136
   human, 154
Position
   changing, 522–523
   definition of, 522
Potato plants, 78
Potential energy, 488, 499, 500
Power plants
   fossil fuel, 508–509
   generators, 492
   geothermal, 501
   hydroelectric, 500
   kilowatt-hours generated by, 504
Power plant technicians, 514
Prairie ecosystem, 132
Precambrian Era, 254
Precipitation
   air mass temperature and, 290–291
   definition of, 268
   formation of, 271
   with fronts, 292
   kinds of, 276–277
   sinkhole formation and, 209
   in water cycle, 268, 269
   weathering and, 208, 210, 211
Predators, 176–177
Prediction, 12
   thermometers, 447
Pressure
   air, 295
   rock formation and, 198
Prey, 176–177
Procyon, 325
Producers, 166, 180
Properties of light, 440
Protists, 32, 33, 38
Protozoans, 38
Pry bars, 555–557
PT-109, 430–431
Puffballs, 46

Pulleys, 562–563
Pupa, 86
Pyramids, 222–223
Pyrite, 195

Q

Quartzite, 204

R

Race horses, 526
Radiation, heat, 452
Rain, 268, 276
   acid, 152
   climate and, 144
Rain forests, 146
   climates of, 144, 145
   layers of, 146
Rain gauges, 296
Rain shadows, 286
Ramps. *See* Inclined planes
Recording data, 16
   flood modeling, 275
   lighting a bulb, 473
Red stars, 325
Red-winged blackbirds, 135
Redwood trees, 40
Reef fossils, 252
Reflection
   of light, 441
   by minerals, 195
   of sound, 426
Refraction (light), 443
Renewable resources, 150
Reproduction, 66, 67, 83
   asexual, 78
   of ferns, 77
   of plants by spores, 44–45, 76–77
Reproductive system (vertebrates), 50
Reptiles, 51
   molting by, 84
   reproduction of, 83

Resources
   nonrenewable, 151
   renewable, 150
Respiratory system, (vertebrates), 50
Rhinoceros, 114
Rhyolite, 196
River hogs, 169
Rivers, 244
   deltas, 235
   valley formation by, 233
Robins, 110
Robotics, 577
Rock(s)
   definition of, 194
   igneous, 196, 204
   magma, 240
   metamorphic, 198, 204
   minerals in, 194, 198
   sedimentary, 197, 204, 250
   in soil horizons, 217
   weathering of, 208–210
   See also Rock cycle
Rock climbing, 560
Rock cycle, 202–204
Roller coasters, 539
Roots, 442
Ross 154, 325
Rovers (spacecraft), 2, 18
Rulers, 4
Runoff, 272

S

Saber-toothed cats, 118
Sac fungi, 46
Safety
   in science investigations, R36
   during severe weather, 280
Sagebrush, 174
Sailboats, 562–563
Salamanders, 51
Salmon, 108
Salt, solubility of, 361
Salt water, 359, 360
Sand, 218

Sand, solubility of, 361
Sand bars, 246
Sand dunes, 235, 246
Sandpiper, 109
Sandstone, 192, 197, 204, 218
Saturn, 314, 318
Savanna climate, 144, 145
Scales, 8
Science projects
   book balancing, 579
   change of reference, 547
   changes in physical properties, 401
   colors in leaves, 59
   conductors, 465
   decomposition, 123
   deep freeze, 225
   density, 367
   earthworm behavior, 123
   fruit fly life cycle, 91
   generating electricity, 515
   insulators and conductors, 515
   linked traits, 91
   long-distance listening, 433
   making rain gauges, 301
   mass of air, 401
   mold growth, 59
   motion and frames of reference, 547
   pulleys, 579
   rivers and landform changes, 259
   rock cycle signs, 225
   seeing around corners, 465
   seismic waves, 259
   sound over distance, 433
   spiral galaxies, 333
   sundials, 333
   temperature and particle movement, 367
   weather and seasons, 301
Scientific inquiries. *See* Inquiries
Scientific method, 20–22
Scorpions, 53
Screws, 568, 572–573

Sea anemones, 29, 52
Sea arches, 246
Sea breezes, 284
Sea horses, 83
Seasons, Earth's tilt and, 308–309
Sea turtles, 63
Second-level consumers, 178
Sediment, 212
   erosion of, 212
   in soil, 216
Sedimentary rocks, 197, 204
   fossils in, 250
Seedling stage, 74
Seeds, 74–75
Seesaws, 552, 556
Segway Human Transporter, 576
Seismograms, 243
Sequence, R22–R23
   hoisting books, 561
Series circuits, 478, 479
Shadows, 440
   rain shadows, 286
Shale, 197, 198
Shape, 346
Shauer, Evan, 88–89
Shelter, need for, 99
Shield volcanoes, 242
Shrimps, 53
SI. *See* International System
Sidewinders, 174, 175
Silt, 197, 218
Silver, 379
Simple machines, 555
   inclined planes, 570–571
   levers, 556–558
   pulleys, 562–563
   screws, 568, 572–573
   wedges, 574
   wheel-and-axles, 564–565
Sinkholes, 209
Sirens, weather, 280
Sirius, 325
Skeletal system, R6–R7
Skeleton races, 544–545

Index

Skunk vine, 141
Skunks, 30
Slate, 198
Sleet, 268, 276–277
Slot canyons, 229
Slugs, 48, 52
Smith, Celia, 156–157
Snails, 52
Snakes, 120–121
   molting by, 84
Snow, 268, 276–278
Socha, Jake, 120–121
Soil
   climate and, 144
   colors of, 219, 224
   contour plowing and, 148
   in ecosystems, 142
   erosion of, 152, 153
   in floodplains, 234
   formation of, 214, 216–217
   horizons of, 217–219
   nutrients in, 220
   plants and, 220
   pollution of, 152
   types of, 218–219
   weathering and, 208
Solar cells, 502
Solar energy, 456–457, 502
   evaporation and, 270
   water cycle and, 268
Solar flares, 324–325
Solar system, 316
   asteroids, 321
   comets, 321
   definition of, 316
   inner planets, 317
   objects in, 316
   outer planets, 318–319
   sun, 324
Solar Two (California), 456
Solar year, 312
Solids, 384, 385
   changes in state of, 354
   definition of, 350
   density of, 345
   measuring, 5
   mixtures of, 358
   in solution with liquids, 360
Solstices, 309
Solubility, 361
Solutions, 360
SONAR, 431
Soule, Chris, 544
Sound
   hearing, 424. *See also* Hearing
   intensity, 411–412
   loudness, 419
   mosquitoes and, 432
   pitch, 410, 418
   reflection of, 426
   sources of, 408–409
   transmission of, 428
Sound waves, 409, 411, 416–420
   absorption of, 427
   amplitude, 419
   frequency, 418
   path of, 424
   wavelength, 417
Southern Hemisphere
   constellations visible in, 328
   seasons in, 309
Sow bugs, 170
Space, lack of sound in, 424
Space shuttle, 390
Species, 66
Speed(s), 524–525
   acceleration and, 531
   comparing, 526
   definition of, 524
   of fastest tennis serve, 528
   on highways, 520
   velocity and, 530
Sperm cells
   human, 67
   plant, 75
Sphinx moths, 86
Spiders, 53
   growth and development of, 84
   instincts of, 106
   molting by, 84
Spiral galaxies, 326, 327
Spirochetes, 37
Sponges (animals), 52
Spores, plant reproduction by, 44–45, 76–77
Sporophyte generation, 77
Spring equinox, 309
Spring scales, 8
Squall lines, 282
Standard measure, 4
Stars, 324–325
   colors of, 325
   definition of, 324
   formation of, 325
   groups of, 326–327
   seasonal positions of, 328
   visibility of, 305
States of matter, 346, 350–351
   changes in, 352, 354, 384–385
Static electricity, 474–475
Stationary fronts, 292
Steam engines, 459
Steel, 380
Stems (plants), 43, 442
Stentor, 32
Stirrup (ear), 425, 428
Stone Forest (China), 191
Storm surges, 278
Storms
   ice, 288
   safety during, 280
   sea breeze, 285
   types of, 278–279
   water pollution from, 152
Stratus clouds, 293
Strawberry plants, 78
Streak (minerals), 195
Streptobacillus, 32
Strike-slip faults, 243
Sugar, solubility of, 361
Sukumar, Raman, 184
Sulfur, 380
Summarizing (in reading), R24–R25
Summer solstice, 309

Sun, 324
   force of, 538
   heat from, 454, 456
   solar energy from, 502
   in solar system, 316
Sunlight
   climate and, 144
   collecting, 454
   in ecosystems, 142
Sunspots, 324
Supporting details, R16–R17
Suspensions, 362
Switches, electric, 482
Systems (human body)
   circulatory, R4–R5
   digestive, R2–R3
   muscular, R8–R9
   skeletal, R6–R7
Systems (nature). *See* Ecosystems

Tables, reading, R28
Taiga ecosystems, 136, 144, 145
Takakai, 76
Talons, 129
Tapirs, 169
Tarantulas, 52, 174
Taylor, Chip, 88–89
Tea, willow bark, 151
Teeth, adaptations of, 100
Telstra Stadium (Sydney, Australia), 10
Temperate forest climate, 144, 145
Temperate rain forest climate, 144
Temperature
   of air over land vs. water, 284
   in Earth's core, 240
   heat and, 448–449
   on mountains, 232
   states of matter and, 352
   weathering and, 210
   on weather maps, 294, 295
Texture, 346
Thermal energy, 449
Thermograms, 446
Thermometers, 8, 448
Thermostats, 457
Thomas Farm (Florida), 112
Thorax (insects), 54
Three states of matter, 384
Thunderstorms, 278, 282, 292
Tigers, 98, 110
Time
   geologic time scale, 254
   speed and, 524–525
Time/space relationships, 14
Titanic, 430
Tools, 2–7
   for measuring, 2–5
   for observing and handling, 6–7
Top-level consumers, 178
Topography, 234
Topsoil, 214, 217
Tornadoes, 278, 280
Toucans, 96
Trace fossils, 251
Traits, 66, 68–69
   acquired, 68, 69
   definition of, 66
   inherited, 68, 69
   variations in, 70
Translucent materials, 442
Transmission
   definition of, 428
   of sound, 428
Transparent materials, 442
Transverse waves, 416, 417
Tree ferns, 33
Trees
   fossil, 251
   fossil vs. present-day, 117
   growth and decay of, 102
   identifying, 69
   redwood, 40
Triceratops, 114
Trilobites, 252, 254
Tropical rain forest climates, 144, 145
Tsunamis, 257
Tubers, 78
Tulips, 70
Tundra ecosystems
   climate of, 144, 145
   diversity in, 146
Turbines, 471, 492, 500, 509
Turtles, 114
   hibernation, 107

Umbra, 332
Understory (rain forests), 146
Universe, 326
Upper mantle, (Earth) 240
Upper St. Croix River gorge (Wisconsin), 230
Uranus, 318, 319
Ursa Major, 326
Using numbers, 13
   hoisting books, 561
   ruler vibrations, 415

V

Vacuoles, 35
Valleys, 233
   rivers and formation of, 244
Vampire bats, 100
Vascular plants, 42–43
VCRs (video cassette recorders), 494
Vehicles, top speeds of, 526
Veins (leaves), 43
Velocity, 530, 531
Venus, 317
Vertebrates, 50–51
Vibrations
   definition of, 408
   of guitar strings, 414

Index

size of object and, 410
sound, 408–409, 428
Video cassette recorders (VCRs), 494
Villa-Komaroff, Lydia, 122
Virtual training, 545
Vision, R10
light and, 444
Volcanic rock, 196
Volcanoes, 242
Cumbre Vieja, 256–257
islands created by, 236, 241
lava from, 196, 200
Mount St. Helens, 258
studying, 258
Volume, 344
measuring, 5
metric system measurements, R33

W

Waning moon, 310
Ward, Steven, 257
Warm air masses, 290, 291
Warm fronts, 292, 294
Waste heat, 460
Water
as basic need, 99
changes in state of, 384–386, 396
densities of, 340
in ecosystems, 142
erosion and, 212
formation of, 394
forms of, 271
groundwater, 272
in oceans, 266
pollution of, 152
rivers, 244
runoff, 272
in soil, 216, 220
states of, 352, 353
in water cycle, 268–269
waves of, 210, 212, 416, 417

See also Water cycle
weathering and, 208, 210, 211
See also Precipitation
Water cycle, 268–269
kinds of precipitation in, 276
landforms and, 284–286
parts of, 270–271
weather related to. *See* Weather
Waterfalls, 500
Water faucets, 564–565
Waterlilies, 134
Water pollution, 152
Water vapor, 268, 270
Wavelength, 417
Waves
landform changes from, 246
light, 426
longitudinal, 417
ocean, 210, 212
sound, 409, 411, 416–420
transverse, 416, 417
water, 210, 212, 246, 416, 417
Waxing moon, 310
Weather
air masses, 290–291
cloud types, 293
fronts, 292
land breezes, 284
measuring, 296
precipitation, 276–277
rain shadows, 286
safety considerations, 280
sea breezes, 284
sea breeze storms, 285
storms, 278–279
Weathering, rock, 208–210
Weather maps, 294–295
Weather stations, 295, 296
Weaverbirds, 106
Wedges, 574
Weight, 540
mass vs., 540
Wetlands, 154
Whales, 104
migration of, 108

songs of, 104, 422
Wheat, 151
Wheel-and-axles, 564–565
Wheelbarrow, 555, 556
White Sands (New Mexico), 235
Wichtowski, Blake, 158
Willow bark tea, 151
Wind, 246
erosion and, 212
in hurricanes, 278
landform changes from, 246
weathering and, 210, 211
on weather maps, 295
Wind farms, 496
Wind tunnels, 18, 546
Wind vanes, 296
Winter solstice, 309
Woodchucks, 107
Wood, petrified, 116, 251
Woolly mammoths, 118, 251
Work, 554–555. *See also* Machines
World Ice Art Championships (Alaska), 348
Worms, 52
Wyoming, fossils in, 253

Y

Years, 312
on planets, 319
Yeast, 46
Yellow stars, 325
Yellowstone National Park, 38, 134
Yukon, 265

Z

Zebras, 106, 144
Zoo veterinarians, 90

Photo Credits

KEY: (*t*) top, (*b*) bottom, (*l*) left, (*r*) right, (*c*) center, (*bg*) background, (*fg*) foreground

Cover
(front) Alaska Stock Images; (back) Art Wolfe/The Image Bank/Getty Images; (back) (*bg*) Tom Walker/Visuals Unlimited

Front End Sheets
Page 1 Bruce Lichtenberger/Peter Arnold, Inc.; **Page 2** (*t*) Bruce Lichtenberger/Peter Arnold, Inc.; (*b*) Ray Coleman/Visuals Unlimited; (*bg*) Jim Steinberg.Photo Researchers; **Page 3** (*t*) Tom Brakefield/The Image Works/Getty Images; (*b*) Gerard Lacz/Peter Arnold

Title Page
Alaska Stock Images

Copyright Page
(*bg*) Tom Walker/Visuals Unlimited; (inset) Alaska Stock Images

Back End Sheets
Page 1 Yva Momatiuk/John Eastcott/Minden Pictures; (*b*) Tom Walker/Visuals Unlimited; (*bg*) Jim Steinberg.Photo Researchers; **Page 2** (*t*) T. Kitchin/V. Hurst/Photo Researchers; (*c*) Jim Brandenburg/Minden Pictures; (*b*) Frieder Blickle/Peter Arnold, Inc.; (*bg*) Jim Steinberg.Photo Researchers; **Page 3** (*l*) Klein/Peter Arnold, Inc.; (*r*) Tim Fitzharris/Minden Pictures; (*bg*) Jim Steinberg.Photo Researchers

Table of Contents
v Stuart Westmoreland/CORBIS; **vii** David Muench/Corbis; **ix** Ted Kinsman/Photo Researchers

Introduction
x–1 (*c*) S Frink/Masterfile; **2** (*c*) AP Photo/NASA; **7** (inset) Sinclair Stammers/Science Photo Library; **10** (*c*) AP Photo/David J.Phillip; **18** (*c*) National Research Council Canada; **20** (inset) Mark Gibson/Index Stock Imagery, Inc.; **20** (*bc*) Robert Llewellyn/CORBIS;

Unit A
26 Reuters/CORBIS; **27** Mark Newman/Bruce Coleman, Inc.; **28–29** Stuart Westmoreland/CORBIS; **30** (*c*) Robert Winslow/Animals Animals; **32** (*r*) BSIP Agency/Index Stock Imagery; (*l*) Eric V. Grave/Photo Researchers; (*bg*) Garry Black/Masterfile; **33** (*r*) Bill Beatty/Visuals Unlimited; (*c*) Wally Eberhart/Visuals Unlimited; (*b*) Adam Jones/Visuals Unlimited; **34** (*l*) Biophoto Associates/Photo Researchers; **35** (*r*) LSHTM/Photo Researchers; **36** (*tl*) Jan Hinsch/Photo Researchers; (*cl*) Biophoto Associates/Photo Researchers; (*bg*) Astrid & Hanns-Frieder Michler/Photo Researchers; **37** (*cl*) Tom Adams/Visuals Unlimited; (*cr*) Science Photo Library/Photo Researchers; (*bl*) Dr. Kari Lounatmaa/Photo Researchers; **38** (*tr*) M.I. Walker/Photo Researchers; (*cl*) Microfield Scientific LTD/Photo Researchers; **40** (*bg*) Freeman Patterson/Masterfile; (*c*) Dan Suzio/Photo Researchers; **42** (*br*) Steve Satushek/Getty Images; **44** (inset) Norman Owen Tomalin/Bruce Coleman; (*bg*) David Noton/Masterfile; **45** (*tr*) James Richardson/Visuals Unlimited; (*tl*) Henry Robison/Visuals Unlimited; **46** (*tr*) Robert Pickett/CORBIS; (*tcr*) SciMAT/Photo Researchers; (*cr*) Carolina Biological/Visuals Unlimited; (*bc*) Jacqui Hurst/CORBIS; (*br*) E. R. Degginger/Color-Pic; (*bg*) CORBIS; **48** Roger Archibald/Animals Animals; **50** (*bl*) Frans Lanting/Minden Pictures; (*br*) Ken Lucas/Ardea London; **51** (*tl*) Michael & Patricia Fogden/Minden Pictures; **52** (*tl*) OSF/Mantis W.F./Animals Animals; (*cl*) Paul Sutherland/Inedependent Photography Network; (*cl*) Alex Fradkin/Images.com/Independent Photography Network; **53** (*tl*) Lightwave Photography/Animals Animals; (*cl*) Roger de la Harpe/Animals Animals; **54** (*r*) Pascal Goetgheluck/Ardea; **56** (*t*) James H Robinson/Photo Researchers; (*b*) Getty Images; **57** (*t*) Getty Images; (inset) Getty; **58** (*t*) JamesRobinson/Photo Researchers; (*b*) Roy Morsch/Corbis; **59** (*bg*) CORBIS; **62–63** Bill Curtsinger/Getty Images; **64** John Daniels/Ardea; **66** (*b*) Getty Images; **67** (*t*) Rob Lewine/CORBIS; (*cr*) Getty Images; **68** (*tl*) Ray Kachatorian/Getty Images; (*tr*) Gio Barto/Getty Images; (*bl*) AGE Fotostock; (*bc*) H. & D. Zielske/Peter Arnold, Inc.; (*br*) Jonathan Nourok/PhotoEdit; **69** (*tl*) Phil Degginger/Animals Animlas; (*tcl*) Klein/Hubert/Peter Arnold; (*tc*) Burke/Triolo Productions/Getty Images; (*tcr*) Jack Milchanowski/Visuals Unlimited; (*tr*) Frans Lanting/Minden Pictures; (*cr*) Arthur Morris/Visuals Unlimited; **70** (*tr*) E. A. Janes/AGE Fotostock; (*cr*) Bruce Coleman, Inc.; (*bl*) John Anderson/Animals Animals; (*br*) Cheryl Ertlet/Visuals Unlimited; **72** Dwight Kuhn; **74** (*bl*) Brent Bergherm/AGE Fotostock; (*bc*) Rachel Weill/Getty Images; (*br*) Carltons/Getty Images; **76** (*cl*) Michael Gadomski/Photo Researchers; (*br*), (*cr*) Dwight Kuhn; **77** (*br*) Steve Satushek/AGE Fotostock; **78** (*l*) Santiago Fernandez/AGE Fotostock; (*r*), (*cl*) Alamy Images; (*cr*) Dwight Kuhn; (*bg*) Bernard Photo Productions/Animals Animals; **80** Ray Coleman/Visuals Unlimited; **82** (*tr*) SuperStock; (*bl*) Georgette Douwma/Getty Images; (*br*) Zigmund Leszczynski/Animals Animals; **83** (*l*) Curtis Richter/Alamy Images; (*r*) Morales/AGE Fotostock; **84** (*r*) Cosmos Blank/Photo Researchers; (*cl*) Norbert Rosing/Getty Images; (*bl*) Tom Brakefield/CORBIS; **85** (*tl*) Susan Solie Patterson/CORBIS; (*tr*) David Zelick/Getty Images; (*cl*) Photodisc/Getty Images; (*c*) Photodisc/Getty Images; (*cr*) Hans-George Gaul/Getty Images; **86** (*cl*), (*cr*) & (*bl*) Dwight Kuhn; (*br*) Patti Murray/Animals Animals; **91** (*bg*) USDA/Science Source/Photo Researchers; **94–95** Dave Watts/Nature Picture Library; **96** Luiz C. Marigo/Peter Arnold; **98** Peter Arnold/Peter Arnold, Inc.; **99** (*cr*) E & P Bauer/Bruce Coleman, Inc.; (*br*) Gil Lopez Espina/Visuals Unlimited; **99** (*cl*) Masa Ushioda/Bruce Coleman, Inc.; **100** (*cl*) Jean Paul Ferrero/Ardea London; (*cr*) Lynn Stone/Animals Animals; (*bl*) Michael & Patricia Fogden/Minden Pictures; (*br*) David Moore/Alamy Images; **101** (*t*) Chase Swift/CORBIS; (*tcr*) Alan G. Nelson/Animals Animals; **102** (*cl*) Brad Mitchell/Alamy Images; (*b*) Robert W. Domm/Visuals Unlimted; **104** Francois Gohier/Ardea; **106** (*bl*) Bernard Castelein/Nature Picture Library; (*bcr*) Hanne & Jens Eriksen/Nature Picture Library; **107** (*tr*) Patti Murray/Animals Animals; (*tcr*) Georgette Douwma/Nature Picture Library; (*cr*) Dietmar Nill/Nature Picture Library; (*bl*) Jennifer Loomis/Animals Animals; (*br*) Doug Wechsler/Animals Animals; **108** (*tr*) Nigel Bean/Nature Picture Library; (*br*) Bob Cranston/Animals Animals; **109** (*tr*) Staffan Widstrand/Nature Picture Library; (*bcr*) Arthur Morris/CORBIS; **110** (*tr*) Ray Richardson/Animals Animals; (*c*) Michael Habicht/Animals Animals; (*bl*) M. Watson/Ardea; **112** Tammy L. Johnson/Florida Museum of Natural History; **114** (*bg*) Mitsuaki Iwago/Minden Images; (*cr*) Paul A. Souders/CORBIS; (*br*) Roger Harris/Science Photo Library; **115** (*t*)(*bg*) SuperStock; (*tl*) Jeff Gage/Florida Museum of Natural History; (*cr*) Peter Scoones/Science Photo Library; (*br*) Dr. Schwimmer/Bruce Coleman, Inc.; **116** (*bg*) Carol Havens/Corbis; (*br*) Kevin Schafer/CORBIS; **117** (*tl*) E.R. Degginger/Bruce Coleman, Inc.; (*tc*) Barry Runk/Stan/Grant Heilman Photography; (*tr*) Ken Lucas/Visuals Unlimited; (*bl*) Patti Murray/Animals Animals; (*bcl*) D. Robert & Lorri Franz/CORBIS; (*br*) David Cavagnaro/Visuals Unlimited; **118** (*cl*) Gianni Dagli Orti/CORBIS; (*cr*) Hulton Archive/Getty Images; (*b*) Ron Testa/The Field Museum; **121** (*tr*) Luiz C. Marigo/Peter Arnold; **122** (*t*) Whitehead Institute for Biomedical Research; (*b*) Photo Researchers; **123** (*bg*) Dwight Kuhn; **125** (*tl*) Hal Brindley/VWPICS/Alamy Images.

Unit B
126–127 Joann Whitmore; **127** AP/Wide World Photos; **128–129** Fritz Polking/Visuals Unlimited; **130** Bob Thomas/Getty Images; **132** (*cr*) Bob & Clara Calhoun/Bruce Coleman; (*br*) Beth Davidow/Visuals Unlimited; **132** (*b*) Cathy Melloan/PhotoEdit; **133** (*t*) E. R. Degginger/Bruce Coleman, Inc.; (*cl*) George Sanker/Bruce Coleman, Inc.; **134** (*t*) Dennis MacDonald/PhotoEdit; (*tl*) Dennis MacDonald/PhotoEdit; **135** (*bl*) Steve Maslowski/Visuals Unlimited; (*b*) Kenneth Fink/Bruce Coleman, Inc.; **136** (*cl*) Jeremy Woodhous/Getty Images; **138** Karl Kummels/SuperStock; **140** (*c*) Jerome Wexler/Visuals Unlimited; (*bl*) Adam Jones/Visuals Unlimited; (*br*) Rob Simpson/Visuals Unlimited; **141** David L. Shirk/Animals Animals; **145** (*tr*) Greg Neise/Visuals Unlimited; (*tl*) Julie Eggers/Bruce Coleman, Inc.; (*cl*) Adam Jones/Visuals Unlimited; (*cr*) Richard Thom/Visuals Unlimited; (*bl*) Patrick Endres/Visuals Unlimited; (*br*) Eastcott-Momatiuk/The Image Works; **148** Jim Richardson/CORBIS; **150** (*bl*) Dennis Brack/IPN; (*b*) Joel Sartore/Getty Images; (*tl*) Micheal Rose/Photo Researchers; (*tr*) Stuart Westmorland/CORBIS; (*cl*) CORBIS; (*cr*) Masterfile; **152** (*tcr*) Susan Van Etten/Photo Edit; (*r*) Mark Richards/PhotoEdit; (*bcr*) Steve Maslowski/Visuals Unlimited; (*br*) Rick Poley/Visuals Unlimited; **153** (*tl*) Cary Wolinsky/IPN; (*tr*) Nancy Richmond/The Image Works; **154** (*cr*) Frank Ordonez/Syracuse Newspapers/The Image Works; (*bl*) CORBIS; **156–157** All Photos NOAA; **158** (*bg*)Index Stock; (inset) Courtesy Mutual of Omaha; **159** (*bg*) Dennis MacDonald/Alamy Images; **161** (*b*) Eastcott-Momatiuk/The Image Works; **162–163** Tom and Pat Leeson; **164** Kim Taylor/Bruce Coleman, Inc.; (*tl*) Darrell Gulin/CORBIS; (*tr*) Lynn Stone/Animals Animals; (*br*) D. Robert & Lorri Franz/CORBIS; **169** (*tc*) Lynn Stone/Animals Animals; (*cl*) Kevin Schafer/CORBIS; (*cr*) Bob Barber/Barber Nature Photography; (*bl*) Michael Fogden/Animals Animals; **170** (*cr*) Wolfgang Kaehler/CORBIS; (*cl*) Ken Lucas/Visuals Unlimited; (*b*) Jim Brandenburg/Minden Picures; **171** (*br*) D.Hurst/Alamy Images; **172** CDC/PHIL/CORBIS; **174** (*cr*) Gerry Ellis/Minden Pictures; (*b*) Royalty-Free/CORBIS; (*br*) Darrell Gulin/CORBIS; **174** (*tr*) ZSSD/MINDEN PICTURES; **175** (*tl*) CORBIS; (*tr*) Dale Sanders/Masterfile; (*c*) Andrew J. Martinez/Photo Researchers; (*br*) Dale Sanders/Masterfile; **176** (*bl*) M. Timothy O'Keefe/Bruce Coleman, Inc.; (*br*) Doug Perrine/SeaPics.com; (*bg*) Wolfgang Kaehler/CORBIS; **177** (*tl*) Jim Zipp/Photo Researchers; (*tc*) Bill Brooks/Masterfile; (*cr*) Barry Runk/Stan/Grant Heilman Photography; (*bl*) Joe McDonald/CORBIS; **182** (*bl*) Animals Animals; **183** (*t*) Wildlife Conservation Society; (*br*) Animals Animals; **184** (*t*) Assign Nature Conservation Foundation; (*b*) Corbis; **185** (*bg*) Mark Mattock/Getty Images

Unit C
188–189 Larry Crumpler/New Mexico Museum of Natural History and Science; **189** John Elk III; **190–191** Keren Su/CORBIS; **192** Kerrick James/Getty Images; **194** (*bl*) Herve Berthoule/JACANA/Science Photo Library; (*bc*) Mark A. Schneider/Visuals Unlimited; (*br*) Bildagentur/Alamy Images; **195** (*tl*) Arnold Fisher/Science Photo Library; (*cl*) Marli Miller/Visuals Unlimited; **195** (*tr*) Marli Miller/Visuals Unlimited; (*cr*) Michael Barnett/Science Photo Library; **196** (*cl*), (*tcr*), (*cr*), & (*bl*) Wally Eberhart/Visuals Unlimited; (*br*) Sciencephotos/Alamy Images; **197** (*tl*) Glenn Oliver/Visuals Unlimited; (*tr*) E. R. Degginger/Color-Pic; (*cl*) Joyce Photographics/Photo Researchers; (*cr*) Wally Eberhart/Visuals Unlimited; **198** (*tcr*) A. J. Copley/Visuals Unlimited; (*bcr*) & (*bl*) Wally Eberhart/Visuals Unlimited; (*br*) Joyce Photographics/Photo Researchers; **200** Ronen Zilberman/AP Photo; **204** (*tr*) Ed Degginger/Color-Pic; (*bl*) Albert J. Copley/Visuals Unlimited; **206–207** (*tr*) John Lawrence/Getty Images; **208** (*bcr*) Peter Kubal/Pan Photo; (*b*) Claire Selby/Animals Animals; **209** (*bg*) AP PHOTO; **210** (*cl*) Freeman Patterson/Masterfile; (*b*) George and Monserrate Schwartz/Alamy Images; **211** (*t*) Masterfile; (*cr*) John Kieffer/Peter Arnold, Inc.; **212** (*r*) Jim Wark/Airphoto; (*bl*) Greg Probst; **214** Louie Psihoyos/IPN; **218** (*b*) Wally Eberhart/Visuals Unlimited; **219** (*tc*) Roland Liptak/Alamy Images(*tr*) Peter Griffith/Masterfile; **219** (*tl*) Jim Craigmyle/CORBIS; (*fg*) Diane Hirsch/Fundamental Photographs; **220** (*bl*) Nigel Cattlin/Alamy Images; (*br*) Jeff Morgan/Alamy Images; **221** (*br*) Dorothea Lange/Library of Congress; **222** Roger Wood/Corbis; **222–223** Stone/Getty Images; **224** (*t*) Photographer's Choice/Getty Images; (*b*) AP/Wide World Photos; **225** (*bg*) Tom Vezo/Peter Arnold, Inc.; **228–229** Mediacolor's/Alamy Images; **230** G. Alan Nelson Outdoor Photography; **232** (*bl*) Paul A. Souders/Corbis; (*br*) Darrell Gulin/Corbis; **233** (inset) Gary Yeowell/Getty Images; (*bg*) Josef Beck/Getty Images; **234** (*bg*) James Strachan/Getty Images; (inset) AirphotoNA.com; **235** (*t*) NASA/Corbis, **235** (*b*) David Muench/Corbis; **236** (*cr*) Photri/Topham/The Image Works; (*b*) Douglas Peebles; **238** Jean-Paul Ferrero/Ardea; **241** (*bg*) David Paterson/Getty Images; **242** (*bg*) Bernhard Edmaier/Science Photo Library; **243** (*tr*) David Butow/Corbis SABA; (*br*) David Hume Kennerly/Getty Images; **244** (*tr*) Ernest Manewal/Index Stock Imagery; (*br*) Pat O'Hara/CORBIS; **245** (*tr*) Arnulf Husmo/Getty Images; (*tl*) Harvey Lloyd/Getty Images; **246** (*bg*) M. T. O'Keefe/Robertstock.com; (inset) Mark Gibson/Index Stock Imagery; **248** James L. Amos/Corbis; **250** (*bg*) Runk/Schoenberger/Grant Heilman Photography, Inc.; **251** (*tl*) Francois Gohier/Photo Researchers; (*tr*) Eberhard Grames/Bilderberg/Peter Arnold, Inc.; (*c*) Imagenes de Nuestro Mundo; **252** (inset) Indiana Dept. of Natural Resources/Falls of the Ohio State Park; (*b*) Indiana Dept. of Natural Resources/Falls of the Ohio River; **253** (*cr*) Sinclair Stammers/Photo Researchers; **259** (*bg*) James King-Holmes/Science Photo Library; **261** (*bl*) Martin Siepmann/AGE footstock

Unit D
262 The Daily Journal, International Falls, MN.; **262–263** Eddie Brady/Lonely Planet Images; **264–265** SuperStock; **266** Jeff Greenberg/The Image Works; **274** David Sailors/CORBIS; **276** (*b*) Royalty-Free/CORBIS; (*tr*) Oote Boe/Alamy Images; (*cr*) Layne Kennedy/Dembinksy Photo Assoc.; (*b*) Rob Atkins/Getty Images; **278** (*tr*) Royalty-Free/CORBIS; (*bl*) Alaska Stock Images; **279** (*bg*) Royalty-Free/CORBIS; **280** (*cl*) Gene Rhoden; (*cr*) Micheal Heller/911 Pictures; (*bl*) Silver Image; (*br*) Richard Cummins/CORBIS; **282** Alan R. Moller/Getty Images; **285** (*cr*) NOAA/AP Photo; **288** Gene Rhoden/Peter Arnold, Inc.; **293** (*tl*) Eastcott/Momatiuk/The Image Works; (*tr*) Tom Dietrich/Getty Images; (*cl*) John Eastcott & Yva Momatiuk/Getty Images; (*cr*) George Post/Photo Researchers; (*b*) Getty Images; (*tc*) Paul Seheult; (*tcr*) Burke/Triolo Productions/Getty Images; (*c*) David Young-Wolff/Photo Edit; (*bg*) Eye Ubiquitous/CORBIS; **301** (*bg*) Robert Carr/Bruce Coleman, Inc.; **303** (*tl*) Layne Kennedy/Dembinsky Photo Associates; **304–305** David Nunuk/Photo Researchers; **306** Warren Flagler/Index Stock Imagery; **313** (*bl*) Gianni Dagli Orti/CORBIS; **314** NASA/JPL/Space Science Institute; **316** (*bg*) StockTrek/Getty Images; (*bcr*) USGS /Photo Researchers; **317** (*tr*) US Geological Survey/Photo Researchers; (*bcr*) Getty Images; **318** (*bg*) Getty Images; (*br*) & (*cl*) NASA; **319** (*tr*) NASA; (*c*) Getty Images; (*bl*) STScl/NASA/Photo Researchers; **320** (*c*) Jerry Lodriguss/Photo Researchers; (*bl*) Reuters/CORBIS; **322** Reuters/CORBIS; **324** (*bl*) GoodShoot/SuperStock; **326** (*bl*) John Chumack/Photo Researchers; **327** (*bg*) Celestial Image Co./Photo Researchers; (*bl*) NASA; **330–331** All photos NASA; **332** (*t*) Photo Researchers; (*b*) Tony Freeman/Photo Researchers; (inset) Peter Falkner; **333** (*bg*) Frank Zullo/Photo Researchers

Unit E

336 SSPL/The Image Works; 337 AP/Wide World Photos; 338–339 Science Photo Library/Photo Researchers; 340 Steve Shott/DK Images; 344 (*tl*) PhotoDisc/Getty Images; 345 (*tc*) Getty Images; (*tr*) Charles D. Winters/Photo Researchers; 346 (*cr*) Flat Earth/FotoSearch; 348 Patrick J. Endres/AlaskaPhotoGraphics; 350 (*bl*) Charles D. Winters/Photo Researchers; (*br*) Mike Hipple/Index Stock Imagery; 351 (*tl*) David Bishop/FoodPix/Getty Images; (*tc*) Spencer Jones/FoodPix/Getty Images; 352 (*br*) Keate/Masterfile; 353 (*bl*) Keate/Masterfile; (*br*) Japack Company/CORBIS; 354 (*bg*) Charles O'Rear/CORBIS; (*cl*) Charles D. Winters/Photo Researchers; 356 David Whitten/Index Stock Imagery/PictureQuest; 361 (*bg*) Peter French/Bruce Coleman; 362 (*bg*) Tim Fitzharris/Minden Pictures; 364 AP/Wide World Photos; 364–365 AP/Wide World Photos; 366 (*t*) AP/Wide World Photos; (*b*) Peter Beck/Corbis; 367 (*bg*) SuperStock/PictureQuest; 369 (*bc*) Daryl Benson/Masterfile; (*br*) John Warden/Index Stock Imagery; 370–371 Gerd Ludwig/VISUM/The Image Works; 377 (*cl*) Richard Megna/Fundamental Photographs; 378 (*tr*) Charles D. Winters/Photo Researchers; (*bl*) Bill Aron/Photo Edit; (*br*) Tony O'Brien/The Image Works; 379 (*tr*) Russ Lappa/Photo Researchers; (*bc*) Jeff J. Daly/Fundamental Photographs; 379 (*c*) Richard Cummins/CORBIS; (*cr*) Tom Pantages/AGPix; (*br*) Ryan McVay/Getty Images; 380 (*tl*) Elio Ciol/CORBIS; (*cr*) Richard Treptow/Photo Researchers; 382 Rommel/Masterfile; 384 (*tcr*) Gary Buss/Getty Images; (*br*) Joseph Van Os/Getty Images; 385 (*cr*) Royalty-Free/CORBIS; 386 (*l*) Kjell Sandved/Bruce Coleman; (*br*) Michael Dalton/Fundamental Photographs; 387 (*tl*) PhotoDisc/Getty Images; (*cr*) Bonnie Kamin/PhotoEdit; 390 NASA/Science Photo Library; 392 (*tc*) Scott Haag; (*bc*) Tom Pantages/AGPix; (*b*) Andrew Lambert Photography/Science Photo Library/Photo Researchers; 394 (*tr*) SuperStock; (*br*) Fife/Photo Researchers; 395 (*br*) Richard Megna/Fundamental Photographs; 396 (*tr*) Grant Heilman/Grant Heilman Photography; (*cr*) Grant Heilman/Grant Heilman Photography; (*bc*) Burke/Triolo/Brand X Pictures/PictureQuest; 400 (*t*) Daum photos; (*b*) Layne Kennedy/Corbis; 401 (*bg*) Royalty-Free/CORBIS; 404–405 Jean Pragen/Getty Images; 406 David Schmidt/Masterfile; (*cl*) Rob Lacey/Vividstock/Alamy Images; (*cr*) Jeremy Woodhouse/Getty Images; (*b*) CORBIS; 411 (*tl*) Iain Davidson Photographic/Alamy Images; (*tr*) Ted Kimsman/Photo Researchers; 412 (*tc*) Alan Schein Photography/CORBIS; (*bl*) John Foxx/Alamy Images; 414 CORBIS; 416 (*bl*) Japack Company/CORBIS; 418 (*cl*) Jeremy Woodhouse/Getty Images; (*bl*) Rob Lacey/Vividstock/Alamy Images; 420 (*cr*) Russell Underwood/CORBIS; (*bl*) James Noble/CORBIS; 422 Tim Davis/CORBIS; 427 (*t*) Roger Ressmeyer/CORBIS; 430–431 The Everett Collection.; 430 (inset) Newscom; 432 (*tl*), (*cl*), (*br*) Julie Wolf Alissi; (*bl*) Will Crocker/Getty Images; 436–437 Maximillian Stock LTD/Science Photo Library; 438 Loren Winters/Visuals Unlimited; 441 (*tr*) Robert Harding Picture Library LTD/Alamy Images; (*cr*) Rolf Bruderer/Masterfile; 442 (*cr*) Grant Heilman Photography; 443 (*tr*) & (*tl*) Richard Megna/Fundamental Photographers; 446 Ted Kinsman/Photo Researchers; 448 (*bl*) Grant Heilman Photography; 449 (*cr*) Peter Bowater/Photo Researchers; (*b*) Big Cheese Photo/PictureQuest; 450 (*r*) Grant Heilman Photography; 451 (*r*) Clouds Hill Imaging LTD/Science Photo Library; 452 (*tr*) David Michael Zimmerman/CORBIS; (*bl*) Stephen J. Krasemann/Photo Researchers; *br*) Ted Kinsman/Photo Researchers; 545 Paul Almasy/CORBIS; 456 (*b*) CORBIS; 457 (*r*) IPS Agency/Index Stock Imagery; (*cl*) Michael Keller/Index Stock Imagery; 458 (*cl*) Rachel Weill/FoodPix; 459 (*tr*) R. Ian Lloyd/Masterfile; 460 (*tr*) Lon C. Diehl/PhotoEdit; (*b*) Pasieka/Science Photo Library; 460 (*cr*) Randy Faris/CORBIS; 462–463 Hans Strand/Corbis; 464 Ray Carson/UF News+Public Affairs; 465 (*bg*) Tim Thompson/CORBIS; 467 (*tcl*) Randy Faris/CORBIS

Unit F

468 Courtesy of Discovery World - The James Lovell Museum of Science Economics and Technology - Milwaukee, WI; 469 LMR Group/Alamy; 470–471 Mark Gibson/Visuals Unlimted; 472 Robert Essel NYC/CORBIS; 474 (*b*) Science Museum, London/Topham-HIP/The Image Works; 480 (*tl*) Andrew Lambert Photography/Photo Researchers; (*cr*) Image Farm Inc./Alamy Images; 481 (*br*) Loren Winters/Visuals Unlimited; 484 Bruce Gray; 486 (*br*) Peter Arnold/Peter Arnold, Inc.; 487 (*bg*) Digital Vision/Getty Images; (*tl*) Tony Freeman/Photo Edit; 488 (*cr*) & (*cl*) Richard Megna/Fundamental Photographs; 489 (*tl*) Biodisc/Visuals Unlimited; 490 (*b*) Tony Freeman/Photo Edit; 491 (*t*) Jeremy Walker/Photo Researchers; 491 (*b*) Jack Plekan/Fundamental Photographs; 492 (*br*) Tom Pantages; 494 (*tc*) David Young-Wolff/Photo Edit; 494 (*c*) Robert Mathena/Fundamental Photographs; 496 Martin Miller/Visuals Unlimited; 500 (*bl*) Jeff Greenberg/Visuals Unlimited; 501 (*t*) Martin Bond/Science Photo Library; 502 (*c*) Martin Bond/Peter Arnold, Inc.; (*bl*) Pegasus/Visuals Unlimited; 504 Angelo Cavalli/Getty Images; 506 (*bl*) Digital Vision/AGE Fotostock; 507 (*tc*) H. Spichtinger/Masterfile; 509 (*tr*) Maximillian Stock LTD/Phototake; (*tl*) Kawasaki Motors Corporation USA; 510 (*tr*) Felicia Martinez/Photo Edit; (*cr*) Anne Dowie/Index Stock Imagery; (*bl*) Peter Arnold/Peter Arnold, Inc.; 512 Courtesy SRI International; 514 (*t*) Queens Borough Library; (*b*) Stone/Getty; 515 (*bg*) Thinkstock/Getty Images; 516 (*cr*) Acestock LTD/Alamy Images; 518–519 Adastra/Getty Images; 520 Macduff Everton/CORBIS; 522 (*b*) Jim Cummins/CORBIS; 524 (*bc*) C Squared Studios/Getty Images; (*b*) David Young-Wolff/PhotoEdit; 525 (*tl*) Leslie O'Shaushnessy/Visuals Unlimited; (*tr*) Spencer Grant/PhotoEdit; 526 (*br*), (*bcl*) Dennis MacDonald/PhotoEdit; 528 Lori Adamski Peek/Getty Images; 530 (*b*) Jim Craigmyle/CORBIS; (*br*) CORBIS; 531 (*tr*) Bill Bachmann/PhotoEdit; (*tcl*) Jonathan Nourok/PhotoEdit; 532 (*tcl*) Wally McNamee/CORBIS; (*tr*) Chris Trotman/Duomo/CORBIS; 533 (*tc*) & (*tcl*) Lon C. Diehl/PhotoEdit; 536 Tom Sanders/CORBIS; 539 (*b*) Bill Bachmann/PhotoEdit; 540 (*tr*) Roy Morsch/Bruce Coleman; (*b*) Jose Luis Pelaez, Inc./CORBIS; 542 (*cl*) Kelly-Mooney Photography/CORBIS; (*bl*) Richard Hutchings/PhotoEdit; (*bg*) Michael Ferguson; 544 Matthew Stockman/Getty Images; 545 Robert LaBerge/Getty Images; 546 GRC/NASA; 547 (*bg*) James Pickerell/The Image Works; 550–551 Ron Stroud/Masterfile; 552 Rune Hellestad/CORBIS; 555 (*t*) Grant Heilman Photography; 556 (*r*) William Sallaz/Duomo/CORBIS; 557 (*tl*) Cindy Charles/Photo Edit; 558 (*c*) Lon C. Diehl/Photo Edit; (*bl*) Zefa/Masterfile; 562 (*b*) Farley Lewis/Photo Researchers; 563 (*tr*) Ted Spiegel/CORBIS; (*br*) Philip Gould/CORBIS; 568 Will Funk/Alpine Aperture; 570 (*b*) Richard Cummins/CORBIS; 571 (*tr*) Tony Freeman/Photo Edit; (*bcl*) David Young-Wolff/Photo Edit; (*br*) Lori Adamski Peek/Getty Images; 573 *tr*) Image Farm/PictureQuest; (*cl*) Tom Schierlitz/Getty Images; 574 (*bl*) Marshall Gordon/Cole Group/Getty Images; 576 Zuma/Newscom; 577 (*cr*)AP/Wide World Photos, (*bl*) AP/Wide; 579 (*bg*) Brad Wright/Brad Wright, Inc.; 580 (*br*) Sky Bonillo/PhotoEdit; 581 K.Hackenberg/Masterfile

Health Handbook

R5 Dennis Kunkel/Phototake; R12 (*t*) CNRI/Science Photo Library/Photo Researchers; R12 (*tc*) A. Pasieka/Photo Researchers; R12 (*bc*) CNRI/Science Photo Library/Photo Researchers; R12 (*b*) Custom Medical Stock Photo; R15 (inset) David Young-Wolff/PhotoEdit; R15 (*b*) Bill O'Connor/Peter Arnold, Inc.

All other photos © Harcourt School Publishers. Harcourt Photos provided by the Harcourt Index, Harcourt IPR, and Harcourt photographers; Weronica Ankarorn, Victoria Bowen, Eric Camden, Doug Dukane, Ken Kinzie, April Riehm, and Steve Williams.